ADVANCED
DC/AC
INVERTERS

APPLICATIONS IN RENEWABLE ENERGY

Power Electronics, Electrical Engineering, Energy, and Nanotechnology Series

Fang Lin Luo and **Hong Ye**, Series Editors
Nayang Technological University, Singapore

PUBLISHED TITLES

Advanced DC/AC Inverters: Applications in Renewable Energy
Fang Lin Luo and Hong Ye

ADVANCED
DC/AC
INVERTERS

APPLICATIONS IN RENEWABLE ENERGY

Fang Lin Luo
Hong Ye

CRC Press
Taylor & Francis Group
Boca Raton London New York

CRC Press is an imprint of the
Taylor & Francis Group, an **informa** business

MATLAB® is a trademark of The MathWorks, Inc. and is used with permission. The MathWorks does not warrant the accuracy of the text or exercises in this book. This book's use or discussion of MAT-LAB® software or related products does not constitute endorsement or sponsorship by The MathWorks of a particular pedagogical approach or particular use of the MATLAB® software.

CRC Press
Taylor & Francis Group
6000 Broken Sound Parkway NW, Suite 300
Boca Raton, FL 33487-2742

© 2013 by Taylor & Francis Group, LLC
CRC Press is an imprint of Taylor & Francis Group, an Informa business

First issued in paperback 2017

No claim to original U.S. Government works
Version Date: 20121031

ISBN 13: 978-1-138-07284-8 (pbk)
ISBN 13: 978-1-4665-1135-4 (hbk)

Visit the Taylor & Francis Web site at
http://www.taylorandfrancis.com

and the CRC Press Web site at
http://www.crcpress.com

Contents

Preface

This book provides knowledge and applications of advanced DC/AC inverters that are both concise and useful for engineering students and practicing professionals. It is well organized in about 300-plus pages and with 250 diagrams to introduce more than 100 topologies of the advanced inverters originally developed by the authors. Some cutting-edge topologies published recently are also illustrated in this book. All prototypes are novel approaches and great contributions to DC/AC inversion technology.

DC/AC inversion technology is one of the main branches in power electronics. It was established in the 1960s and grew fast in the 1980s. DC/AC inverters convert DC power sources to AC power users. It is of vital importance for all industrial applications, including electrical vehicles and renewable energy systems. In recent years, inversion technology has been rapidly developed and new topologies have been published, which largely improved the power factor and increased the power efficiency. One purpose of writing this book is to summarize the features of DC/AC inverters and introduce more than 50 new circuits as well.

DC/AC Inverters can be sorted into two groups: pulse-width modulation (PWM) inverters and multilevel modulation (MLM) inverters. People are familiar with PWM inverters, such as the voltage source inverter (VSI) and current source inverter (CSI). They are very popular in industrial applications. The impedance-source inverter (ZSI) was first introduced in 2003 and immediately attracted many experts of power electronics to this area. Its advantages are so attractive for research and industrial applications that hundreds of papers regarding ZSI have been published in recent years.

All PWM inverters have the same main power circuits, that is, three legs for three-phase output voltage. Multilevel inverters were invented in the 1980s. Unlike PWM inverters, multilevel inverters have different main power circuits. Typical ones are the diode-clamped inverters, capacitor clamped (flying capacitor) inverters, and hybrid H-bridge multilevel inverters. Multilevel inverters overcame the drawbacks of the PWM inverter and opened a broad way for industrial applications.

This book introduces four novel multilevel inverters proposed by the authors: laddered multilevel inverters, super-lift modulated inverters, switched-capacitor inverters, and switched-inductor inverters. They have simple structures with fewer components to implement the DC/AC inversion. They are very attractive to DC/AC inverter designers and have been applied in industrial applications, including renewable energy systems.

This book introduces four methods to manage the switching angles to obtain the lowest THD, which is an important topic for multilevel inverters. The half-height (HH) method is superior to others in achieving low THD by careful investigation. A MATLAB® program is used to search the best switching angles to obtain the lowest THD. The best switching angles for any multilevel inverter are listed in tables as convenient references for electrical engineers. Simulation waveforms are shown to verify the design.

Due to world energy resource shortage, the development of renewable energy sources is critical. The relevant topics such as energy-saving and power supply quality are also paid much attention. Renewable energy systems require large number of DC/DC converters and DC/AC inverters. In this book, introduction and design examples including analysis and results are given for wind turbine and solar panel energy systems.

The book is organized in 15 chapters. General knowledge is introduced in Chapter 1. Traditional PWM inverters, such as voltage source inverters, current source inverters, and impedance source inverters, are discussed in Chapters 2 to 5. New quasi-impedance source inverters and soft-switching PWM inverters are investigated in Chapters 6 and 7, respectively. Multi-level DC/AC inverters are generally introduced in Chapter 8. Trinary H-bridge inverters are specially investigated in Chapter 9. Novel multilevel inverters including laddered multilevel inverters, super-lift modulated inverters, switched capacitor inverters, and switched inductor inverters are introduced in Chapters 10 to 13. Best switching angles to obtain lowest THD for multilevel DC/AC inverters are studied in Chapter 14. Application examples in renewable energy systems are discussed in Chapter 15.

Professor Fang Lin Luo
AnHui University
HeFei, China

Doctor Hong Ye
Nanyang Technological University
Singapore

Authors

Dr. Fang Lin Luo is a full professor at AnHui University, China. He also has a joint appointment at Nanyang Technological University Singapore. He was an associate professor in the School of Electrical and Electronic Engineering, Nanyang Technological University (NTU), Singapore in 1995–2012. He received his BSc degree, first class, with honors (magna cum laude) in radio-electronic physics at the Sichuan University, Chengdu, China, and his PhD in electrical engineering and computer science (EE and CS) at Cambridge University, England, in 1986.

After his graduation from Sichuan University, he joined the Chinese Automation Research Institute of Metallurgy (CARIM), Beijing, China, as a senior engineer. From there, he then went to the Entreprises Saunier Duval, Paris, France, as a project engineer in 1981–1982. He worked with Hocking NDT Ltd., Allen-Bradley IAP Ltd., and Simplatroll Ltd. in England as a senior engineer after he earned his PhD from Cambridge. He is a fellow of Cambridge Philosophical Society and a senior member of IEEE. He has published 13 books and 300 technical papers in IEE/IET proceedings and IEEE transactions, and various international conferences. His present research interest focuses on power electronics and DC and AC motor drives with computerized artificial intelligent control (AIC) and digital signal processing (DSP), and AC/DC and DC/DC and AC/AC converters and DC/AC inverters, renewable energy systems, and electrical vehicles.

He is currently associate editor of *IEEE Transactions on Power Electronics* and associate editor of *IEEE Transactions on Industrial Electronics*. He is also the international editor of *Advanced Technology of Electrical Engineering and Energy*. Dr. Luo was chief editor of *Power Supply Technologies and Applications* from 1998 to 2003. He was the general chairman of the first IEEE Conference on Industrial Electronics and Applications (ICIEA 2006) and the third IEEE Conference on Industrial Electronics and Applications (ICIEA 2008).

Dr. Hong Ye is a research fellow with the School of Biological Sciences, Nanyang Techological University, Singapore. She received her bachelor's degree, first class, in 1995; her master's degree in engineering from Xi'an Jiaotong University, China, in 1999; and a PhD degree from Nanyang Technological University (NTU), Singapore, in 2005.

She was with the R&D Institute, XIYI Company, Ltd., China, as a research engineer from 1995 to 1997. She worked at NTU as a research associate from 2003 to 2004 and has been a research fellow from 2005.

Dr. Ye is an IEEE member and has coauthored 13 books. She has published more than 80 technical papers in IEEE transactions, IEE proceedings, and other international journals, as well as presenting them at various international conferences. Her research interests are power electronics and conversion technologies, signal processing, operations research, and structural biology.

1

Introduction

DC/AC inverters convert DC source energy for AC users, and are a big category of power electronics. Power electronics is the technology to process and control the flow of electric energy by supplying voltages and currents in a form that is optimally suited for user loads [1]. A typical block diagram is shown in Figure 1.1 [2]. The input power can be AC and DC sources. A general example is that the AC input power is from the electric utility. The output power to load can be AC and DC voltages. The power processor in the block diagram is usually called a *converter*. Conversion technologies are used to construct converters. Therefore, there are four categories of converters [3]:

- AC/DC converters/rectifiers (AC to DC)
- DC/DC converters (DC to DC)
- DC/AC inverters/converters (DC to AC)
- AC/AC converters (AC to AC)

We will use converter as a generic term to refer to a single power conversion stage that may perform any of the functions listed above. To be more specific, in AC to DC and DC to AC conversion, *rectifier* refers to a converter when the average power flow is from the AC to the DC side. *Inverter* refers to the converter when the average power flow is from the DC to the AC side. In fact, the power flow through the converter may be reversible. In that case, as shown in Figure 1.2 [2], we refer to that converter in terms of its rectifier and inverter modes of operation.

1.1 Symbols and Factors Used in This Book

We list the factors and symbols used in this book here. If no specific description is given, the parameters follow the meaning stated here.

1.1.1 Symbols Used in Power Systems

For instantaneous values of variables such as voltage, current, and power that are functions of time, the symbols used are lowercase letters v, i, and p,

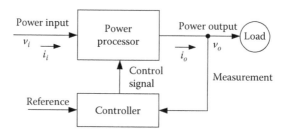

FIGURE 1.1
The block diagram of a power electronics system.

respectively. They are functions of time operating in the time domain. We may or may not explicitly show that they are functions of time, for example, using v rather than $v(t)$. The uppercase symbols V and I refer to their average value in DC quantities and a root-mean-square (rms) value in AC quantities, computed from their instantaneous waveforms. They generally refer to an average value in DC quantities and a root-mean-square (rms) value in AC quantities. If there is a possibility of confusion, the subscript *avg* or *rms* is used. The average power is always indicated by P.

Usually, the input voltage and current are represented by v_{in} and i_{in} (or v_1 and i_1), and the output voltage and current are represented by v_O and i_O (or v_2 and i_2). The input and output powers are represented by P_{in} and P_O. The power transfer efficiency (η) is defined as $\eta = P_O/P_{in}$.

Passive loads such as resistor R, inductor L, and capacitor C are generally used in circuits. We use R, L, and C to indicate their symbols and values as well. All these parameters and their combination Z are linear loads since the performance of the circuit constructed by these components is described by a linear differential equation. Z is the impedance of a linear load. If the circuit consists of a resistor R, an inductor L, and a capacitor C connected in series, the impedance Z is represented by

$$Z = R + j\omega L - j\frac{1}{\omega C} = |Z| \angle \phi \tag{1.1}$$

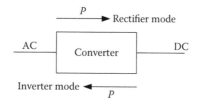

FIGURE 1.2
AC-to-DC converters.

where R is the resistance measured by Ω, L is the inductance measured by H, C is the capacitance measured by F, ω is the AC supply angular frequency measured by rad/s, and $\omega = 2\pi f$, where f is the AC supply frequency measured by Hz. For the calculation of Z, if there is no capacitor in the circuit, the term $j\frac{1}{\omega C}$ is omitted (do not take $c = 0$ and $j\frac{1}{\omega C} => \infty$). The absolute impedance $|Z|$ and the phase angle ϕ are determined by

$$|Z| = \sqrt{R^2 + \left(\omega L - \frac{1}{\omega C}\right)^2}$$

$$\phi = \tan^{-1}\frac{\omega L - \frac{1}{\omega C}}{R}$$

(1.2)

Example 1.1

A circuit has a load with a resistor $R = 20\ \Omega$, an inductor $L = 20$ mH, and a capacitor $C = 200\ \mu F$ in series connection. The voltage supplying frequency $f = 60$ Hz. Calculate the load impedance and the phase angle.

Solution:

From Equation (1.1), the impedance Z is

$$Z = R + j\omega L - j\frac{1}{\omega C} = 20 + j120\pi \times 0.02 - j\frac{1}{120\pi \times 0.0002}$$

$$= 20 + j(7.54 - 13.26) = 20 - j5.72 = |Z|\angle\phi$$

From Equation (1.2), the absolute impedance $|Z|$ and phase angle ϕ are

$$|Z| = \sqrt{R^2 + (\omega L - \frac{1}{\omega C})^2} = \sqrt{20^2 + 5.72^2} = 20.8\ \Omega$$

$$\phi = \tan^{-1}\frac{\omega L - \frac{1}{\omega C}}{R} = \tan^{-1}\frac{-5.72}{20} = -17.73°$$

If a circuit consists of a resistor R and an inductor L connected in series, the corresponding impedance Z is represented by

$$Z = R + j\omega L = |Z|\angle\phi$$

(1.3)

The absolute impedance $|Z|$ and phase angle ϕ are determined by

$$|Z| = \sqrt{R^2 + (\omega L)^2}$$

$$\phi = \tan^{-1} \frac{\omega L}{R}$$

(1.4)

We define the circuit time constant τ as

$$\tau = \frac{L}{R}$$

(1.5)

If a circuit consists of a resistor R and a capacitor C connected in series, the impedance Z is represented by

$$Z = R - j\frac{1}{\omega C} = |Z| \angle \phi$$

(1.6)

The absolute impedance $|Z|$ and phase angle ϕ are determined by

$$|Z| = \sqrt{R^2 + \left(\frac{1}{\omega C}\right)^2}$$

$$\phi = -\tan^{-1} \frac{1}{\omega C R}$$

(1.7)

We define the circuit time constant τ as

$$\tau = RC$$

(1.8)

Summary of the Symbols

Symbol	Explanation (Measuring Unit)
C	capacitor/capacitance (F)
f	frequency (Hz)
i, I	instantaneous current, average/rms current (A)
L	inductor/inductance (H)
R	resistor/resistance (Ω)
p, P	instantaneous power, rated/real power (W)
q, Q	instantaneous reactive power, rated reactive power (VAR)
s, S	instantaneous apparent power, rated apparent power (VA)
v, V	instantaneous voltage, average/rms voltage (V)
Z	impedance (Ω)
ϕ	phase angle (degree, or radian)
η	efficiency (percents%)
τ	time constant (second)
ω	angular frequency (radian/sec), $\omega = 2\pi f$

1.1.2 Factors and Symbols Used in AC Power Systems

The input AC voltage can be single-phase or three-phase voltages. They are usually a pure sinusoidal wave function. For a single-phase input voltage $v(t)$, the function can be expressed as [4]:

$$v(t) = \sqrt{2}V \sin \omega t = V_m \sin \omega t \tag{1.9}$$

where v is the instantaneous input voltage, V is its root-mean-square (rms) value, V_m is its amplitude, ω is the angular frequency, $\omega = 2\pi f$, and f is the supply frequency. Usually, the input current may not be a pure sinusoidal wave, depending on the load. If the input voltage supplies a linear load (resistive, inductive, capacitive loads, or their combination) the input current $i(t)$ is not distorted, but may be delayed in a phase angle ϕ. In this case, it can be expressed as

$$i(t) = \sqrt{2}I \sin(\omega t - \phi) = I_m \sin(\omega t - \phi) \tag{1.10}$$

where i is the instantaneous input current, I is its root-mean-square value, I_m is its amplitude, and ϕ is the phase-delay angle. We define the power factor (PF) as

$$PF = \cos \phi \tag{1.11}$$

PF is the ratio of the real power (P) to the apparent power (S). We have the relation $S = P + jQ$, where Q is the reactive power. The power vector diagram is shown in Figure 1.3. We have the following relations between the powers:

$$S = VI^* = \frac{V^2}{Z^*} = P + jQ = |S| \angle \phi \tag{1.12}$$

$$|S| = \sqrt{P^2 + Q^2} \tag{1.13}$$

$$\phi = \tan^{-1} \frac{Q}{P} \tag{1.14}$$

$$P = S\cos \phi \tag{1.15}$$

$$Q = S\sin \phi \tag{1.16}$$

If the input current is distorted, it consists of harmonics. Its fundamental harmonic can be expressed as

$$i_1 = \sqrt{2}I_1 \sin(\omega t - \phi_1) = I_{m1} \sin(\omega t - \phi_1) \tag{1.17}$$

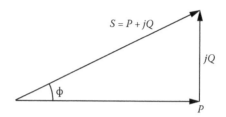

FIGURE 1.3
Power vector diagram.

where i_1 is the fundamental harmonic instantaneous value, I_1 its rms value, I_{m1} its amplitude, and ϕ_1 its phase angle. In this case, the displacement power factor (DPF) is defined as

$$DPF = \cos\phi_1 \tag{1.18}$$

Correspondingly, the power factor is defined as

$$PF = \frac{DPF}{\sqrt{1+THD^2}} \tag{1.19}$$

where THD is the total harmonic distortion. It can be used to measure both voltage and current waveforms. It is defined as

$$THD = \frac{\sqrt{\Sigma_{n=2}^{\infty} I_n^2}}{I_1} \quad \text{or} \quad THD = \frac{\sqrt{\Sigma_{n=2}^{\infty} V_n^2}}{V_1} \tag{1.20}$$

where I_n or V_n is the amplitude of the nth order harmonic.

The harmonic factor (HF) is a variable that describes the weighted percentage of the nth order harmonic with reference to the amplitude of the fundamental harmonic V_1. It is defined as

$$HF_n = \frac{I_n}{I_1} \quad \text{or} \quad HF_n = \frac{V_n}{V_1} \tag{1.21}$$

$n = 1$ corresponds to the fundamental harmonic. Therefore, $HF_1 = 1$. The total harmonic distortion (THD) can be written as

$$THD = \sqrt{\sum_{n=2}^{\infty} HF_n^2} \tag{1.22}$$

A pure sinusoidal waveform has THD = 0.

Weighted total harmonic distortion (WTHD) is a variable to describe waveform distortion. It is defined as follows:

$$WTHD = \frac{\sqrt{\sum_{n=2}^{\infty} \frac{V_n^2}{n}}}{V_1} \tag{1.23}$$

Note that THD gives an immediate measure of the inverter output voltage waveform distortion. WTHD is often interpreted as the normalized current ripple expected in an inductive load when fed from the inverter output voltage.

Example 1.2:

A load with a resistor $R = 20\ \Omega$, an inductor $L = 20$ mH, and a capacitor $C = 200\ \mu F$ in series connection is supplied by an AC voltage of 240 V (rms) with frequency $f = 60$ Hz. Calculate the circuit current and the corresponding apparent power S, real power P, reactive power Q, and the power factor PF.

Solution:

From Example 1.1, the impedance Z is

$$Z = R + j\omega L - j\frac{1}{\omega C} = 20 + j120\pi \times 0.02 - j\frac{1}{120\pi \times 0.0002}$$

$$= 20 + j(7.54 - 13.26) = 20 - j5.72 = 20.8\angle -17.73°\ \Omega$$

The circuit current I is

$$I = \frac{V}{Z} = \frac{240}{20.8\angle -17.73°} = 11.54\angle 17.73°\ A$$

The apparent power S is

$$S = VI^* = 240 \times 11.54\angle -17.73° = 2769.23\angle -17.73°\ VA$$

The real power P is

$$P = |S|\cos\phi = 2769.23 \times \cos 17.73° = 2637.7W$$

The reactive power Q is

$$Q = |S|\sin\phi = 2769.23 \times \sin -17.73° = -843.3VAR$$

The power factor is

$$PF = \cos \phi = 0.9525 \; Leading$$

Summary of the Symbols

Symbol	Explanation (Measuring Unit)
DPF	displacement power factor (percent)
HF_n	nth order harmonic factor
i_1, I_1	instantaneous fundamental current, average/rms fundamental current (A)
i_n, I_n	instantaneous nth order harmonic current, average/rms nth order harmonic current (A)
I_m	current amplitude (A)
PF	power factor (leading/lagging percent)
q, Q	instantaneous reactive power, rated reactive power (VAR)
s, S	instantaneous apparent power, rated apparent power (VA)
t	time (second)
THD	total harmonic distortion (percent)
v_1, V_1	instantaneous fundamental voltage, average/rms fundamental voltage (V)
v_n, V_n	instantaneous nth order harmonic voltage, average/rms nth order harmonic voltage (V)
WTHD	weighted total harmonic distortion (percent)
ϕ_1	phase angle of the fundamental harmonic (degree, or radian)

1.1.3 Factors and Symbols Used in DC Power Systems

We define the output DC voltage instantaneous value to be v_d and the average value to be V_d (or V_{d0}) [5]. A pure DC voltage has no ripple; it is then called ripple-free DC voltage. Otherwise, a DC voltage is distorted and consists of a DC component and AC harmonics. Its rms value is V_{d-rms}. For a distorted DC voltage, its rms value V_{d-rms} is constantly higher than its average value V_d. The ripple factor (RF) is defined as

$$RF = \frac{\sqrt{\sum_{n=1}^{\infty} V_n^2}}{V_d} \tag{1.24}$$

where V_n is the nth order harmonic. The form factor (FF) is defined as

$$FF = \frac{V_{d-rms}}{V_d} = \frac{\sqrt{\sum_{n=0}^{\infty} V_n^2}}{V_d} \tag{1.25}$$

where V_0 is the 0th order harmonic; that is, the average component V_d. Therefore, we obtain $FF > 1$, and the relation

$$RF = \sqrt{FF^2 - 1} \tag{1.26}$$

The form factor FF and ripple factor RF are used to describe the quality of a DC waveform (voltage and current parameters). For a pure DC voltage, FF = 1 and RF = 0.

Summary of the Symbols

Symbol	Explanation (Measuring Unit)
FF	form factor (percent)
RF	ripple factor (percent)
v_d, V_d	instantaneous DC voltage, average DC voltage (V)
$V_{d\text{-rms}}$	rms DC voltage (V)
v_n, V_n	instantaneous nth order harmonic voltage, average/rms nth order harmonic voltage (V)

1.2 FFT—Fast Fourier Transform

The FFT [6] is a very versatile method of analyzing waveforms. A periodic function with radian frequency ω can be represented by a series of sinusoidal functions:

$$f(t) = \frac{a_0}{2} + \sum_{n=1}^{\infty} (a_n \cos n\omega t + b_n \sin n\omega t) \tag{1.27}$$

where the Fourier coefficients are

$$a_n = \frac{1}{\pi} \int_0^{2\pi} f(t) \cos(n\omega t) d(\omega t) \quad n = 0, 1, 2, \ldots \infty \tag{1.28}$$

$$b_n = \frac{1}{\pi} \int_0^{2\pi} f(t) \sin(n\omega t) d(\omega t) \quad n = 1, 2, \ldots \infty \tag{1.29}$$

In this case, we call the terms with radian frequency ω the fundamental harmonic and the terms with radian frequency $n\omega$ ($n > 1$) higher order harmonics. If we draw the amplitudes of all harmonics in the frequency domain, we can get the spectrum in individual peaks. The term $a_0/2$ is the DC component.

1.2.1 Central Symmetrical Periodical Function

If the periodic function is a central symmetrical periodic function, all terms with cosine function disappear. The FFT becomes

$$f(t) = \sum_{n=1}^{\infty} b_n \sin n\omega t \tag{1.30}$$

where

$$b_n = \frac{1}{\pi} \int_0^{2\pi} f(t)\sin(n\omega t)d(\omega t) \quad n = 1, 2, \ldots\infty \tag{1.31}$$

We usually call this the odd function. In this case, we call the term with the radian frequency ω the fundamental harmonic, and the terms with the radian frequency $n\omega$ ($n > 1$) higher order harmonics. If we draw the amplitudes of all harmonics in the frequency domain, we can get the spectrum in individual peaks. Since it is an odd function, the DC component is zero.

1.2.2 Axial (Mirror) Symmetrical Periodical Function

If the periodic function is an axial symmetrical periodic function, all terms with sine function disappear. The FFT becomes

$$f(t) = \frac{a_0}{2} + \sum_{n=1}^{\infty} a_n \cos n\omega t \tag{1.32}$$

where $a_0/2$ is the DC component and

$$a_n = \frac{1}{\pi} \int_0^{2\pi} f(t)\cos(n\omega t)d(\omega t) \quad n = 0, 1, 2, \ldots\infty \tag{1.33}$$

The term $a_0/2$ is the DC component. We usually call this function the even function. In this case, we call the term with the radian frequency ω the fundamental harmonic, and the terms with the radian frequency $n\omega$ ($n > 1$) higher-order harmonics. If we draw the amplitudes of all harmonics in the frequency domain, we can get the spectrum in individual peaks. Since it is an even function, the DC component is usually not zero.

1.2.3 Nonperiodic Function

The spectrum of a periodic function in the time domain is a discrete function in the frequency domain. For a nonperiodic function in the time domain, it is possible to represent it by Fourier integration. The spectrum is a continuous function in the frequency domain.

1.2.4 Useful Formulae and Data

Some trigonometric formulae are useful for FFT:

$$\sin^2 x + \cos^2 x = 1 \quad \sin x = \cos\left(\frac{\pi}{2} - x\right)$$

$$\sin x = -\sin(-x) \qquad \sin x = \sin(\pi - x)$$

$$\cos x = \cos(-x) \qquad \cos x = -\cos(\pi - x)$$

$$\frac{d}{dx}\sin x = \cos x \qquad \frac{d}{dx}\cos x = -\sin x$$

$$\int \sin x \, dx = -\cos x \quad \int \cos x \, dx = \sin x$$

$$\sin(x \pm y) = \sin x \cos y \pm \cos x \sin y$$

$$\cos(x \pm y) = \cos x \cos y \mp \sin x \sin y$$

$$\sin 2x = 2\sin x \cos x$$

$$\cos 2x = \cos^2 x - \sin^2 x$$

Some values corresponding to the special angles are usually used:

$$\sin\frac{\pi}{12} = \sin 15° = 0.2588 \qquad \cos\frac{\pi}{12} = \cos 15° = 0.9659$$

$$\sin\frac{\pi}{8} = \sin 22.5° = 0.3827 \qquad \cos\frac{\pi}{8} = \cos 22.5° = 0.9239$$

$$\sin\frac{\pi}{6} = \sin 30° = 0.5 \qquad \cos\frac{\pi}{6} = \cos 30° = \frac{\sqrt{3}}{2} = 0.866$$

$$\sin\frac{\pi}{4} = \sin 45° = \frac{\sqrt{2}}{2} = 0.7071 \qquad \cos\frac{\pi}{4} = \cos 45° = \frac{\sqrt{2}}{2} = 0.7071$$

$$\tan\frac{\pi}{12} = \tan 15° = 0.2679 \qquad \tan\frac{\pi}{8} = \tan 22.5° = 0.4142$$

$$\tan\frac{\pi}{6} = \tan 30° = \frac{\sqrt{3}}{3} = 0.5774 \qquad \tan\frac{\pi}{4} = \tan 45° = 1$$

$$\tan x = \frac{1}{co - \tan x} \qquad \tan x = co - \tan\left(\frac{\pi}{2} - x\right)$$

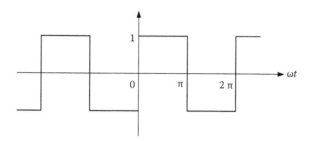

FIGURE 1.4
A waveform.

1.2.5 Examples of FFT Applications

Example 1.3

An odd-square waveform is shown in Figure 1.4. Find the FFT and HF up to
the 7th order, and also the THD and WTHD.

Solution:

The function f(t) is

$$
f(t) = \begin{cases} 1 & 2n\pi \leq \omega t < (2n+1)\pi \\ -1 & (2n+1)\pi \leq \omega t < 2(n+1)\pi \end{cases}
\tag{1.34}
$$

The Fourier coefficients are

$$
b_n = \frac{1}{\pi} \int_0^{2\pi} f(t)\sin(n\omega t)d(\omega t) = \frac{2}{n\pi} \int_0^{n\pi} \sin\theta\, d\theta = 2\frac{1-(-1)^n}{n\pi}
$$

or

$$
b_n = \frac{4}{n\pi} \quad n = 1,3,5,\ldots\infty
\tag{1.35}
$$

Finally, we obtain

$$
F(t) = \frac{4}{\pi} \sum_{n=1}^{\infty} \frac{\sin(n\omega t)}{n} \quad n = 1,3,5,\ldots\infty
\tag{1.36}
$$

The fundamental harmonic has an amplitude of $4/\pi$. If we consider the higher order harmonics until the 7th order, that is, $n = 3, 5, 7$, the HFs are

$$HF_3 = 1/3; \qquad HF_5 = 1/5; \qquad HF_7 = 1/7$$

The *THD* is

$$THD = \frac{\sqrt{\sum_{n=2}^{\infty} V_n^2}}{V_1} = \sqrt{\left(\frac{1}{3}\right)^2 + \left(\frac{1}{5}\right)^2 + \left(\frac{1}{7}\right)^2} = 0.41415 \qquad (1.37)$$

The *WTHD* is

$$WTHD = \frac{\sqrt{\sum_{n=2}^{\infty} \frac{V_n^2}{n}}}{V_1} = \sqrt{\left(\frac{1}{3}\right)^3 + \left(\frac{1}{5}\right)^3 + \left(\frac{1}{7}\right)^3} = 0.219 \qquad (1.38)$$

Example 1.4

An even-square waveform is shown in Figure 1.5. Find the FFT and HF up to the 7th order, and also the *THD* and *WTHD*.
The function $f(s)$ is

$$f(t) = \begin{cases} 1 & (2n - 0.5)\pi \le \omega t < (2n + 0.5)\pi \\ -1 & (2n + 0.5)\pi \le \omega t < (2n + 1.5)\pi \end{cases} \qquad (1.39)$$

The Fourier coefficients are

$$a_0 = 0$$

$$a_n = \frac{1}{\pi} \int_0^{2\pi} f(t)\cos(n\omega t)d(\omega t) = \frac{4}{n\pi} \int_0^{\frac{n\pi}{2}} \cos\theta \, d\theta = \frac{4\sin\frac{n\pi}{2}}{n\pi}$$

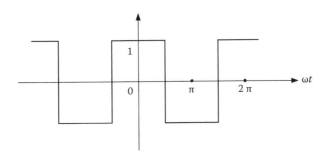

FIGURE 1.5
An even square waveform.

or

$$a_n = \frac{4}{n\pi}\sin\frac{n\pi}{2} \qquad n = 1,3,5,\ldots\infty \tag{1.40}$$

The term $\sin\frac{n\pi}{2}$ is used to define the sign. Finally, we obtain

$$F(t) = \frac{4}{\pi}\sum_{n=1}^{\infty}\sin\frac{n\pi}{2}\cos(n\omega t) \quad n = 1,3,5,\ldots\infty \tag{1.41}$$

The fundamental harmonic has the amplitude $4/\pi$. If we consider the higher order harmonics until the 7th order, that is, n = 3, 5, 7, the HFs are

$$HF_3 = 1/3; \qquad HF_5 = 1/5; \qquad HF_7 = 1/7$$

The *THD* is

$$THD = \frac{\sqrt{\sum_{n=2}^{\infty}V_n^2}}{V_1} = \sqrt{\left(\frac{1}{3}\right)^2 + \left(\frac{1}{5}\right)^2 + \left(\frac{1}{7}\right)^2} = 0.41415 \tag{1.42}$$

The *WTHD* is

$$WTHD = \frac{\sqrt{\sum_{n=2}^{\infty}\frac{V_n^2}{n}}}{V_1} = \sqrt{\left(\frac{1}{3}\right)^3 + \left(\frac{1}{5}\right)^3 + \left(\frac{1}{7}\right)^3} = 0.219 \tag{1.43}$$

Example 1.5

An odd-waveform pulse with pulse width x is shown in Figure 1.6. Find the FFT and HF up to the 7th order, and also the THD and WTHD.

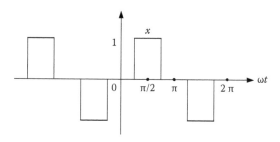

FIGURE 1.6
An odd-waveform pulse.

The function $f(t)$ is in the period $-\pi$ to $+\pi$:

$$f(t) = \begin{cases} 1 & \dfrac{\pi - x}{2} \leq \omega t < \dfrac{\pi + x}{2} \\[3mm] -1 & -\dfrac{\pi + x}{2} \leq \omega t < -\dfrac{\pi - x}{2} \end{cases} \tag{1.44}$$

The Fourier coefficients are

$$b_n = \frac{1}{\pi}\int_0^{2\pi} f(t)\sin(n\omega t)d(\omega t) = \frac{2}{n\pi}\int_{n\frac{\pi-x}{2}}^{n\frac{\pi+x}{2}} \sin\theta\, d\theta = 2\frac{\cos(n\frac{\pi-x}{2}) - \cos(n\frac{\pi-x}{2})}{n\pi}$$

$$= 2\frac{2\cos(n\frac{\pi-x}{2})}{n\pi} = \frac{4\sin(\frac{n\pi}{2})\sin(\frac{nx}{2})}{n\pi}$$

or

$$b_n = \frac{4}{n\pi}\sin\frac{n\pi}{2}\sin\frac{nx}{2} \qquad n = 1,3,5,\ldots\infty \tag{1.45}$$

Finally, we obtain

$$F(t) = \frac{4}{\pi}\sum_{n=1}^{\infty}\frac{\sin(n\omega t)}{n}\sin\frac{n\pi}{2}\sin\frac{nx}{2} \quad n = 1,3,5,\ldots\infty \tag{1.46}$$

The fundamental harmonic has the amplitude $\frac{4}{\pi}\sin\frac{x}{2}$. If we consider the higher order harmonics until the 7th order, that is, n = 3, 5, 7, the HFs are

$$HF_3 = \frac{\sin\frac{3x}{2}}{3\sin\frac{x}{2}}; \quad HF_5 = \frac{\sin\frac{5x}{2}}{5\sin\frac{x}{2}}; \quad HF_7 = \frac{\sin\frac{7x}{2}}{7\sin\frac{x}{2}}$$

The values of the HFs should be absolute.
If $x = \pi$, the THD is

$$THD = \frac{\sqrt{\sum_{n=2}^{\infty}V_n^2}}{V_1} = \sqrt{\left(\frac{1}{3}\right)^2 + \left(\frac{1}{5}\right)^2 + \left(\frac{1}{7}\right)^2} = 0.41415 \tag{1.47}$$

The WTHD is

$$WTHD = \frac{\sqrt{\sum_{n=2}^{\infty}\frac{V_n^2}{n}}}{V_1} = \sqrt{\left(\frac{1}{3}\right)^3 + \left(\frac{1}{5}\right)^3 + \left(\frac{1}{7}\right)^3} = 0.219 \tag{1.48}$$

Example 1.6

A 5-level odd waveform is shown in Figure 1.7. Find the FFT and HF up to the 7th order, and also the THD and WTHD.
  The function $f(t)$ is in the period $-\pi - +\pi$:

$$f(t) = \begin{cases} 2 & \dfrac{\pi}{3} \le \omega t < \dfrac{2\pi}{3} \\[2mm] 1 & \dfrac{\pi}{6} \le \omega t < \dfrac{\pi}{3}, \dfrac{2\pi}{3} \le \omega t < \dfrac{5\pi}{6} \\[2mm] 0 & other \\[2mm] -1 & -\dfrac{5\pi}{6} \le \omega t < -\dfrac{2\pi}{3}, -\dfrac{\pi}{3} \le \omega t < -\dfrac{\pi}{6} \\[2mm] -2 & -\dfrac{2\pi}{3} \le \omega t < -\dfrac{\pi}{3} \end{cases} \qquad (1.49)$$

The Fourier coefficients are

$$b_n = \frac{1}{\pi}\int_0^{2\pi} f(t)\sin(n\omega t)d(\omega t) = \frac{2}{n\pi}\left[\int_{\frac{n\pi}{6}}^{\frac{5n\pi}{6}}\sin\theta\,d\theta + \int_{\frac{n\pi}{3}}^{\frac{2n\pi}{3}}\sin\theta\,d\theta\right]$$

$$= \frac{2}{n\pi}\left[\left(\cos\frac{n\pi}{6} - \cos\frac{5n\pi}{6}\right) + \left(\cos\frac{n\pi}{3} - \cos\frac{2n\pi}{3}\right)\right] = \frac{4}{n\pi}\left(\cos\frac{n\pi}{6} + \cos\frac{n\pi}{3}\right)$$

or

$$b_n = \frac{4}{n\pi}\left(\cos\frac{n\pi}{6} + \cos\frac{n\pi}{3}\right) \qquad n = 1,3,5,\ldots\infty \qquad (1.50)$$

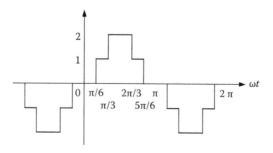

FIGURE 1.7
A five-level odd waveform.

Finally, we obtain

$$F(t) = \frac{4}{\pi} \sum_{n=1}^{\infty} \frac{\sin(n\omega t)}{n} \left(\cos \frac{n\pi}{6} + \cos \frac{n\pi}{3} \right) \quad n = 1, 3, 5, \ldots \infty \quad (1.51)$$

The fundamental harmonic has the amplitude $\frac{2}{\pi}(1+\sqrt{3})$. If we consider the higher-order harmonics until the 7th order, that is, n = 3, 5, 7, the HFs are

$$HF_3 = \frac{2}{3(1+\sqrt{3})} = 0.244; \quad HF_5 = \frac{\sqrt{3}-1}{5(1+\sqrt{3})} = 0.0536; \quad HF_7 = \frac{\sqrt{3}-1}{5(1+\sqrt{3})} = 0.0383$$

The values of the HFs should be absolute.
The THD is

$$THD = \frac{\sqrt{\sum_{n=2}^{\infty} V_n^2}}{V_1} = \sqrt{\sum_{n=2}^{\infty} HF_n^2} = \sqrt{0.244^2 + 0.0536^2 + 0.0383^2} = 0.2527 \quad (1.52)$$

The WTHD is

$$WTHD = \frac{\sqrt{\sum_{n=2}^{\infty} \frac{V_n^2}{n}}}{V_1} = \sqrt{\sum_{n=2}^{\infty} \frac{HF_n^2}{n}} = \sqrt{\frac{0.244^2}{3} + \frac{0.0536^2}{5} + \frac{0.0383^2}{7}} = 0.1436$$

$$(1.53)$$

1.3 DC/AC Inverters

DC/AC inverters [1,2] were not widely used in industrial applications before the 1960s because of their complexity and cost. They were used in most fractional horsepower AC motor drives in the 1970s since AC motors have advantages such as lower cost than DC motors, small size, and they are maintenance-free. Because of advances in semiconductor technology, more effective devices such as IGBTs and MOSFETs were produced in the 1980s, and DC/AC inverters began to be widely applied in industrial applications. Currently, DC/AC conversion techniques can be grouped into two categories: pulse width modulation (PWM) and multilevel modulation (MLM). Each category has many circuits that implement the modulation. Using PWM, we can design various inverters such as voltage source inverters (VSIs), current source inverters (CSIs), impedance source inverters (ZSIs), and multistage PWM inverters.

A single-phase half-wave PWM is shown in Figure 1.8.

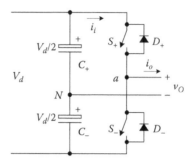

FIGURE 1.8
Single-phase half-wave PWM VSI.

The pulse width modulation (PWM) method is suitable for DC/AC conversion since the input voltage is usually a constant DC voltage (DC link). Pulse phase modulation (PPM) is also possible, but is not so convenient. Pulse amplitude modulation (PAM) is not suitable for DC/AC conversion since the input voltage is usually a constant DC voltage. In PWM operation, all pulses' leading edges start from the beginning of the pulse period, and their trailing edge is adjustable. PWM is the fundamental technique for many types of PWM DC/AC inverters such as VSI, CSI, ZSI, and multistage PWM inverters.

Another group of DC/AC inverters are the multilevel inverters (MLIs). They were invented in the late 1970s. The early MLIs were constructed by diode-clamped and capacitor-clamped circuits. Later, other MLIs were developed.

Three important procedures have to be emphasized in this book:

- To categorize existing inverters
- To introduce updated circuits
- To investigate soft switching methods

1.3.1 Categorizing Existing Inverters

Since the number of inverters is large, we have to sort them systematically. Some circuits have not been precisely named, so their functions cannot be inferred from their names.

1.3.2 Updated Circuits

Many updated DC/AC inverters were developed in recent decades, but not introduced in textbooks. We have to incorporate these techniques in this book and teach students to understand them.

1.3.3 Soft Switching Methods

The soft switching technique has been widely used in switching circuits for a long time. It effectively reduces the power losses of equipment and greatly increases the power transfer efficiency. A few soft switching technique methods will be introduced in this book.

References

1. Luo, F. L. and Ye, H. 2010. *Power Electronics: Advanced Conversion Technologies*, Boca Raton, FL: Taylor & Francis.
2. Luo, F. L., Ye, H., and Rashid, M. H. 2005. *Digital Power Electronics and Applications*. Boston: Academic Press Elsevier.
3. Rashid, M. H. 2004. *Power Electronics: Circuits, Devices and Applications (3rd edition)*. Upper Saddle River, NJ: Prentice Hall.
4. Luo, F. L. and Ye, H. 2007. DC-modulated single-stage power factor correction AC/AC converters. *Proc. ICIEA'2007*, Harbin, China, pp. 1477–1483.
5. Luo, F. L. and Ye, H. 2004. *Advanced DC/DC Converters*. Boca Raton, FL: CRC Press.
6. Carlson A. B. 2000. *Circuits*. Pacific Grove, CA: Brooks/Cole.

2

Pulse Width-Modulated DC/AC Inverters

DC/AC inverters are quickly developed with knowledge of the power switching circuits applied in industrial applications in comparison with other power switching circuits. In the past century, plenty of topologies of DC/AC inverters have been created. DC/AC inverters are mainly used in AC motor adjustable speed drives (ASDs), as shown in Figure 2.1. Power DC/AC inverters have been widely used in other industrial applications since the late 1980s. Semiconductor manufacture development allowed high-power devices such as IGBTs and MOSFETs to operate at higher switching frequencies (e.g., from tens of kHz up to a few MHz). Conversely, some devices such as thyristors (SCRs), GTOs, triacs, and BTs, with lower switching frequency and higher power rate, the IGBT and MOSFET may have both high power rate and high switching frequency [1,5].

Square waveform DC/AC inverters were used well before the 1980s and the thyristor, GTO, and triac could be used in low-frequency switching operations. The power BT and IGBT were produced for high frequency operation. The corresponding equipment implementing the pulsewidth-modulation (PWM) technique has a large range of output voltage and frequency and low THD.

Nowadays, two DC/AC inversion techniques are popular in this area: PWM and MLM. Most DC/AC inverters are still PWM DC/AC inverters in different prototypes. We will introduce PWM inverters in this chapter and MLM inverters in Chapter 8.

2.1 Introduction

DC/AC inverters are used for converting a DC power source into AC power applications. They are generally used in the following applications:

1. Variable voltage/variable frequency AC supplies in adjustable speed drive (ASD), devices such as induction motor drives, synchronous machine drives, and so on

2. Constant regulated voltage AC power supplies, such as uninterruptible power supplies (UPSs)

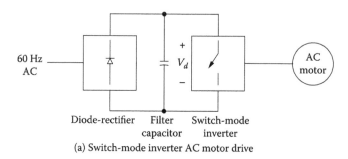

(a) Switch-mode inverter AC motor drive

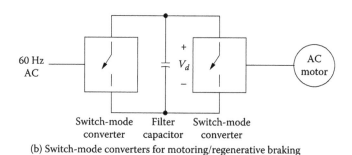

(b) Switch-mode converters for motoring/regenerative braking

FIGURE 2.1
A standard adjustable speed drive (ASD) scheme.

3. Static variability (reactive power) compensations
4. Passive/active series and parallel filters
5. Flexible AC transmission systems (FACTSs)
6. Voltage compensations

Adjustable speed induction motor drive systems are widely applied in industrial applications. These systems require DC/AC power supply with variable frequency usually from 0 Hz to 400 Hz in fractional horsepower (HP) to hundreds of HP. A large number of DC/AC inverters are in the world market. The typical block circuit of an ASD is shown in Figure 2.1. From this block diagram, we can see that the power DC/AC inverter produces variable frequency and voltage to implement ASD.

The PWM technique is different from pulse amplitude modulation (PAM) and pulse phase modulation (PPM). In this technique, all pulses have adjustable width with constant amplitude and phase. The corresponding circuit is called the pulse width modulator. Typical input and output waveforms of a pulse width modulator are shown in Figure 2.2. The output pulse train has

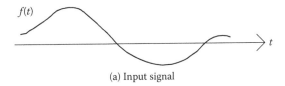

(a) Input signal

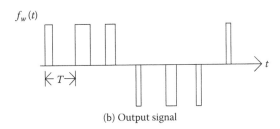

(b) Output signal

FIGURE 2.2
Typical input and output waveforms of a pulse width modulator.

the pulses of the same amplitude and different widths, which corresponds to the input signal at the sampling instants.

2.2 Parameters Used in PWM Operation

Some parameters specially used in PWM operation are introduced in this section.

2.2.1 Modulation Ratios

The modulation ratio is usually obtained from a uniform amplitude triangle (carrier) signal with amplitude V_{tri-m}. The maximum amplitude of the input signal is assumed to be V_{in-m}. We define the amplitude modulation ratio m_a for a single-phase inverter as follows:

$$m_a = \frac{V_{in-m}}{V_{tri-m}} \tag{2.1}$$

We also define the frequency modulation ratio m_f as follows:

$$m_f = \frac{f_{tri-m}}{f_{in-m}} \tag{2.2}$$

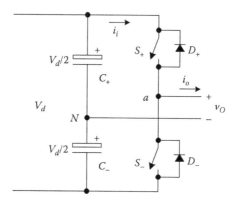

FIGURE 2.3
Single-leg switch-mode inverter.

A single-leg switch-mode inverter is shown in Figure 2.3. The DC-link voltage is V_d. Two large capacitors are used to establish the neutral point N. The AC output voltage from point a to N is V_{AO}, and its fundamental component is $(V_{AO})_1$. We denote $(\hat{V}_{AO})_1$ to show the maximum amplitude of $(V_{AO})_1$. The waveforms of the input (control) signal and triangle signal, and the spectrum of the PWM pulse train are shown in Figure 2.4.

If the maximum amplitude $(\hat{V}_{AO})_1$ of the input signal is smaller than or equal to half of the DC-link voltage $V_d/2$ and the modulation ratio m_a is smaller than or equal to the unity. In this case, the fundamental component $(V_{AO})_1$ of the output AC voltage is proportional to the input voltage. The voltage control by varying m_a for a single-phase PWM is split in three areas, which are shown in Figure 2.5.

2.2.1.1 Linear Range ($m_a \leq 1.0$)

The condition $(\hat{V}_{Ao})_1 = m_a \frac{V_d}{2}$ determines the linear region. It is a sinusoidal PWM where the amplitude of the fundamental frequency voltage varies linearly with the amplitude modulation ratio m_a. The PWM pushes the harmonics into a high-frequency range around the switching frequency and its multiples. However, the maximum available amplitude of the fundamental frequency component may not be as high as desired.

2.2.1.2 Over Modulation ($1.0 < m_a \leq 3.24$)

The condition $\frac{V_d}{2} < (\hat{V}_{Ao})_1 \leq \frac{4}{\pi} \frac{V_d}{2}$ determines the overmodulation region. When the amplitude of the fundamental frequency component in the output voltage increases beyond 1.0, it reaches overmodulation. In the overmodulation range, the amplitude of the fundamental frequency voltage no longer varies linearly with m_a.

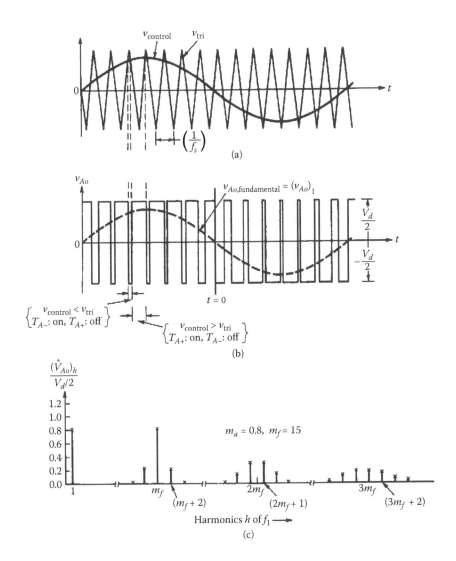

FIGURE 2.4
Pulse width modulation.

Overmodulation causes the output voltage to contain many more harmonics in the sidebands as compared with the linear range. The harmonics with dominant amplitudes in the linear range may not be dominant during overmodulation.

2.2.1.3 Square Wave (Sufficiently Large $m_a > 3.24$)

The condition $(\hat{V}_{Ao})_1 > \frac{4}{\pi}\frac{V_d}{2}$ determines the square wave region. The inverter voltage waveform degenerates from a pulse width modulated waveform into

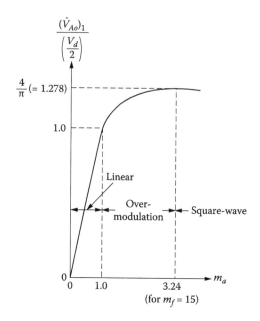

FIGURE 2.5
Voltage control by varying m_a.

a square wave. Each switch of the inverter leg in Figure 2.3 is on for one half-cycle (180°) of the desired output frequency.

2.2.1.4 Small m_f ($m_f \leq 21$)

Usually the triangle waveform frequency is much larger than the input signal frequency to obtain small THD. For the situation with a small $m_f \leq 21$, two points have to be mentioned:

- *Synchronous PWM*. For small values of m_f, the triangle waveform signal and the input signal should be synchronized to each other (synchronous PWM). This synchronous PWM requires that m_f be an integer. Synchronous PWM is used because asynchronous PWM (where m_f is not an integer) results in subharmonics (of the fundamental frequency) that are very undesirable in most applications. This implies that the triangle waveform frequency varies with the desired inverter frequency (e.g., if the inverter output frequency and hence the input signal frequency are 65.42 Hz and $m_f = 15$, the triangle wave frequency should be exactly $15 \times 65.42 = 981.3$ Hz).

- $m_f \leq 21$ *and should be an odd integer*. As discussed previously, m_f should be an odd integer except in single-phase inverters with PWM unipolar voltage switching.

2.2.1.5 Large m_f ($m_f > 21$)

The amplitudes of subharmonics due to asynchronous PWM are small at large values of m_f. Therefore, at large values of m_f, asynchronous PWM can be used where the frequency of the triangle waveform is kept constant, whereas the input signal frequency varies, resulting in nonintegral values of m_f (so long as they are large). However, if the inverter is supplying a load such as an AC motor, the subharmonics at zero or close to zero frequency, even though small in amplitude, will result in large current, which is highly undesirable. Therefore, asynchronous PWM should be avoided.

It is very important to determine the harmonic components of the output voltage. Referring to Figure 2.4c, we have the fast Fourier transform (FFT) spectrum and the harmonics. Choosing the frequency modulation ratio m_f as an odd integer and amplitude modulation ratio $m_a < 1$, we have the generalized harmonics of the output voltage shown in Table 2.1.

The rms voltages of the output voltage harmonics are calculated by the following formula:

$$(V_O)_h = \frac{V_d}{\sqrt{2}} \frac{(\hat{V}_{AO})_h}{V_d/2} \tag{2.3}$$

where $(V_O)_h$ is the hth harmonic rms voltage of the output voltage, V_d is the DC link voltage, and $(\hat{V}_{AO})_h/(V_d/2)$ or $(\hat{V}_{AO})_h/(V_d/2)$ is tabulated as a function of m_a.

TABLE 2.1

Generalized Harmonics of V_O (or V_{AO}) for Large m_f

	m_a				
h	0.2	0.4	0.6	0.8	1.0
1 (Fundamental)	0.2	0.4	0.6	0.8	1.0
m_f	1.242	1.15	1.006	0.818	0.601
$m_f \pm 2$	0.016	0.061	0.131	0.220	0.318
$m_f \pm 4$					0.018
$2m_f \pm 1$	0.190	0.326	0.370	0.314	0.181
$2m_f \pm 3$		0.024	0.071	0.139	0.212
$2m_f \pm 5$				0.013	0.033
$3m_f$	0.335	0.123	0.083	0.171	0.113
$3m_f \pm 2$	0.044	0.139	0.203	0.176	0.062
$3m_f \pm 4$		0.012	0.047	0.104	0.157
$3m_f \pm 6$				0.016	0.044
$4m_f \pm 1$	0.163	0.157	0.008	0.105	0.068
$4m_f \pm 3$	0.012	0.070	0.132	0.115	0.009
$4m_f \pm 5$			0.034	0.084	0.119
$4m_f \pm 7$				0.017	0.050

Note: $(\hat{V}_{AO})_h/(V_d/2)$ or $(\hat{V}_{AO})_h/(V_d/2)$ is tabulated as a function of m_a.

If the input (control) signal is a sinusoidal wave, we usually call this inversion sinusoidal pulse width modulation (SPWM). The typical waveforms of an SPWM are also shown in Figures 2.4a and 2.4b.

Example 2.1

A single-phase half-bridge DC/AC inverter is shown in Figure 2.3 to implement an SPWM with $V_d = 200$ V, $m_a = 0.8$, and $m_f = 27$. The fundamental frequency is 50 Hz. Determine the rms value of the fundamental frequency and some of the harmonics in output voltage using Table 2.1.

Solution:

From Equation (2.3) we have the general rms values

$$(V_O)_h = \frac{V_d}{\sqrt{2}}\frac{(\hat{V}_{AO})_h}{V_d/2} = \frac{200}{\sqrt{2}}\frac{(\hat{V}_{AO})_h}{V_d/2} = 141.42\frac{(\hat{V}_{AO})_h}{V_d/2}V \qquad (2.4)$$

Checking the data from Table 2.1, we can get rms values as follows:

$(V_O)_1 = 141.42 \times 0.8 = 113.14V$ at 50 Hz

$(V_O)_{23} = 141.42 \times 0.818 = 115.68V$ at 1150 Hz

$(V_O)_{25} = 141.42 \times 0.22 = 31.11V$ at 1250 Hz

$(V_O)_{27} = 141.42 \times 0.818 = 115.68V$ at 1350 Hz

$(V_O)_{51} = 141.42 \times 0.139 = 19.66V$ at 2550 Hz

$(V_O)_{53} = 141.42 \times 0.314 = 44.41V$ at 2650 Hz

$(V_O)_{55} = 141.42 \times 0.314 = 44.41V$ at 2750 Hz

$(V_O)_{57} = 141.42 \times 0.139 = 19.66V$ at 2850 Hz

etc.

2.2.2 Harmonic Parameters

Refer to Figure 2.4c, in which various harmonic parameters such as HF_n, THD, and WTHD, which are used in PWM operation, were introduced in Chapter 1.

2.3 Typical PWM Inverters

DC/AC inverters have three typical supply methods:

- Voltage source inverter (VSI)
- Current source inverter (CSI)
- Impedance source inverter (z-source inverter or ZSI)

Generally, the main power circuits of various PWM inverters can be the same. The difference between them is the type of power supply source or network (voltage source, current source, or impedance source).

2.3.1 Voltage Source Inverter (VSI)

A voltage source inverter is supplied by a DC voltage source. In an ASD, the DC source is usually an AC/DC rectifier. There is a large capacitor used to keep the DC-link voltage stable. Usually, a VSI has *buck* operation function. Its output voltage peak value is lower than the DC link voltage.

It is necessary to avoid a *short circuit* across the DC voltage source during the operation. If a VSI operates in bipolar mode, that is, the upper switch and lower switch in a leg work to provide a PWM output waveform, the control circuit and interface have to be designed to leave small gaps between switching signals to the upper switch and lower switch in the same leg. For example, if the output voltage frequency is in the 0–400 Hz range, and the PWM carrying frequency is in the 2–20 kHz range, the gaps are usually set as 20–100 ns. This requirement is not very convenient for the control circuit and interface design. Therefore, unipolar operation is implemented in most industrial applications.

2.3.2 Current Source Inverter (CSI)

A current source inverter is supplied by a DC current source. In an ASD, the DC current source is usually an AC/DC rectifier with a large inductor to keep the supplying current stable. Usually, a CSI has a *boost* operation function. Its output voltage peak value is higher than the DC link voltage.

Since the source is a DC current source, it is necessary to avoid *open circuit* of the inverter during operation. The control circuit and interface have to be designed for small overlaps between switching signals to the upper switches and lower switches at least in one leg. For example, if the output voltage frequency is in the range 0–400 Hz, and the PWM carrying frequency is in the 2–20 kHz range, the overlaps are usually set as 20–100 ns. This requirement is easy to implement for the control circuit and interface design.

2.3.3 Impedance Source Inverter (z-Source Inverter—ZSI)

An impedance source inverter (ZSI) is supplied by a voltage source or current source via an x-shaped impedance network formed by two capacitors and two inductors, which is called a z-network. In an ASD, the DC impedance source is usually an AC/DC rectifier. A z-network is located between the rectifier and the inverter. Since there are two inductors and two capacitors to be set in front of the chopping legs, no restriction to avoid the legs opened or short-circuited. A ZSI has both *buck* and *boost* operation function. Its output voltage peak value can be higher or lower than the DC link voltage.

2.3.4 Circuits of DC/AC Inverters

The generally used DC/AC inverters are as follows:

1. Single-phase half-bridge voltage source inverter (VSI)
2. Single-phase full-bridge VSI
3. Three-phase full-bridge VSI
4. Three-phase full-bridge current source inverter (CSI)
5. Multistage PWM inverter
6. Soft switching inverter
7. Impedance source inverter (ZSI)

References

1. Mohan, N., Undeland, T. M., and Robbins, W. P. 2003. *Power Electronics: Converters, Applications and Design* (3rd edition). New York: John Wiley & Sons.
2. Holtz, J. 1992. Pulsewidth modulation—A survey. *IEEE Trans. Ind. Electron.*, pp. 410–420.
3. Peng, F. Z. 2003. Z-source inverter. *IEEE Trans. Ind. Appl.*, pp. 504–510.
4. Trzynadlowski, A. M. 1998. *Introduction to Modern Power Electronics*. New York: John Wiley & Sons.
5. Anderson, J. and Peng, F. Z. 2008. Four quasi-Z-Source inverters. *Proc. IEEE PESC'2008*, pp. 2743–2749.

3

Voltage Source Inverters

Voltage source inverters (VSIs) are widely used in industrial applications and renewable energy systems [1–3]. Their structure and control circuitry are simple. Therefore, most engineers like to use them. Many semiconductor companies produce VSI control chips to simplify control circuitry design. We introduce some typical VSI circuits in this chapter.

3.1 Single-Phase Voltage Source Inverter

Single-phase voltage source inverters can be implemented in half-bridge and full bridge circuits.

3.1.1 Single-Phase Half-Bridge VSI

A single-phase half-bridge voltage source inverter (VSI) is shown in Figure 3.1. The carrier-based pulse width modulation (PWM) technique is applied in this inverter. Two large capacitors are required to provide a neutral point N; therefore, each capacitor keeps half of the input DC voltage. Since the output voltage is referring to the neutral point N, the maximum output voltage is smaller than half of the DC-link voltage if it is operating in linear modulation. The modulation operations are shown in Figure 2.5. Two switches S+ and S− in one chopping leg are switched by the PWM signal. Two switches S+ and S− operate in an exclusive state with small deadtime to avoid a short-circuit.

In general, linear modulation operation is considered, so that m_a is usually smaller than unity, for example, $m_a = 0.8$. Generally, in order to obtain low THD, the m_f is usually a large number. For description convenience we choose $m_f = 9$. In order to understand each inverter well, we show some typical waveforms in Figure 3.2.

How to determine the pulse width is the key of the PWM. If the control signal v_C is a sine wave function as shown in Figure 3.2a, we call the modulation *sinusoidal pulse width modulation* (SPWM). Figure 3.2b offers the switching signal. When it is "ON" switch on the upper switch S+, and switch off the lower switch S−, vice versa, it is "OFF" switch off the upper switch S+, and switch

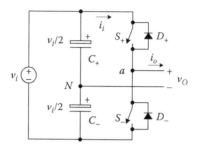

FIGURE 3.1
Single-phase half-bridge VSI. (a) Carrier and modulating signals. (b) Switch S+ state. (c) Switch S− state. (d) AC output voltage. (e) AC output current.

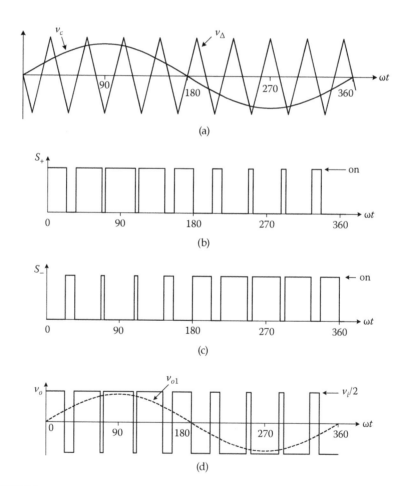

FIGURE 3.2
Single-phase half-bridge VSI ($m_a = 0.8$, $m_f = 9$).

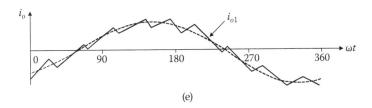

(e)

FIGURE 3.2 (*continued*)
Single-phase half-bridge VSI ($m_a = 0.8$, $m_f = 9$).

on the lower switch S–. Assume that the amplitude of the triangle wave is unity and the amplitude of the sine-wave is 0.8. Referring to Figure 3.2a, the sine wave function is

$$f(t) = m_a \sin \omega t = 0.8 \sin 100\pi t \qquad (3.1)$$

where $\omega = 2\pi f$, $f = 50$ Hz. The triangle functions are lines:

$$f_{\Delta 1}(t) = -4 fm_f t = -1800t \qquad\qquad f_{\Delta 2}(t) = 4 fm_f t - 2 = 1800t - 2$$

$$f_{\Delta 3}(t) = 4 - 4 fm_f t = 4 - 1800t \qquad\qquad f_{\Delta 4}(t) = 4 fm_f t - 6 = 1800t - 6$$

......

$$f_{\Delta(2n-1)}(t) = 4(n-1) - 4 fm_f t \qquad\qquad f_{\Delta 2n}(t) = 4 fm_f t - (4n-2) \qquad (3.2)$$

......

$$f_{\Delta 17}(t) = 32 - 1800t \qquad\qquad f_{\Delta 18}(t) = 1800t - 34$$

$$f_{\Delta 19}(t) = 36 - 1800t$$

Example 3.1

A single-phase half-bridge DC/AC inverter is shown in Figure 3.1 to implement SPWM with $m_a = 0.8$ and $m_f = 9$. Determine the first width of the pulse shown in Figure 3.2a.

Solution:

The leading age of the first pulse is at $t = 0$. Referring to the trigonometric formulae, the first pulse width (time or degree) is determined by the equation

$$0.8\sin 100\pi t = 1800t - 2 \qquad (3.3)$$

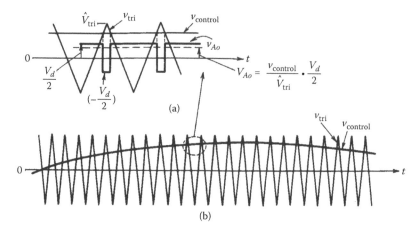

FIGURE 3.3
Sinusoidal PWM.

This is a transcendental equation with the unknown parameter t. Using *iterative method* to solve the equation, let x = 0.8sin100πt and y = 1800t – 2. We can choose the initial t_0 = 1.38889 ms = 25°.

The first pulse width to switch on and off the switch S+ is 1.2861 ms (or 23.15°).

Other pulse widths can be determined from other equations using the iterative method. For a PWM operation with large m_f, readers can refer to Figure 3.3.

Figure 3.2 shows the ideal waveforms associated with the half-bridge VSI. We can find out the phase delay between the output current and voltage. For a large m_f we can see the cross points demonstrated in Figure 3.3 with smaller phase delay between the output current and voltage.

3.1.2 Single-Phase Full-Bridge VSI

A single-phase full-bridge voltage source inverter (VSI) is shown in Figure 3.4. The carrier-based pulse-width modulation (PWM) technique is applied in this inverter. Two large capacitors may be used to provide a neutral point N, but are not necessary. Since the output voltage is not referring to the neutral point N, the maximum output voltage is possibly greater than the half of the DC-link voltage. If it is operating in linear modulation the output voltage is smaller than the DC-link voltage. The modulation operation of multi-leg VSI is different from that of single-leg (single-phase half-bridge); VSI is shown in the previous subsection. It will be shown in Figure 3.8. Four switches, S_1+/S_1- and S_2+/S_2-, in two legs are applied and switched by the PWM signal.

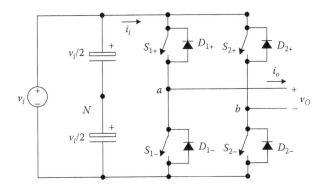

FIGURE 3.4
Single-phase full-bridge VSI. (a) Carrier and modulating signals. (b) Switch S_1+ and S_1- state. (c) Switch S_2+ and S_2- state. (d) AC output voltage. (e) AC output current.

Figure 3.5 shows the ideal waveforms associated with the full-bridge VSI. There are two sine waves used in Figure 3.5a corresponding to the two legs operation. We can find out the phase delay between the output current and voltage.

The method to determine the pulse widths is the same as that introduced in the previous section. Referring to Figure 3.5a, we can find that there are two sine wave functions:

$$f_+(t) = m_a \sin \omega t = 0.8 \sin 100\pi t \tag{3.4}$$

and

$$f_-(t) = -m_a \sin \omega t = -0.8 \sin 100\pi t \tag{3.5}$$

The triangle functions are:

$$f_{\Delta 1}(t) = -4 fm_f t = -1600t \qquad f_{\Delta 2}(t) = 4 fm_f t - 2 = 1600t - 2$$

$$f_{\Delta 3}(t) = 4 - 4 fm_f t = 4 - 1600t \qquad f_{\Delta 4}(t) = 4 fm_f t - 6 = 1600t - 6$$

......

$$f_{\Delta(2n-1)}(t) = 4(n-1) - 4 fm_f t \qquad f_{\Delta 2n}(t) = 4 fm_f t - (4n-2) \tag{3.6}$$

......

$$f_{\Delta 15}(t) = 28 - 1600t \qquad f_{\Delta 16}(t) = 1600t - 30$$

$$f_{\Delta 17}(t) = 32 - 1600t$$

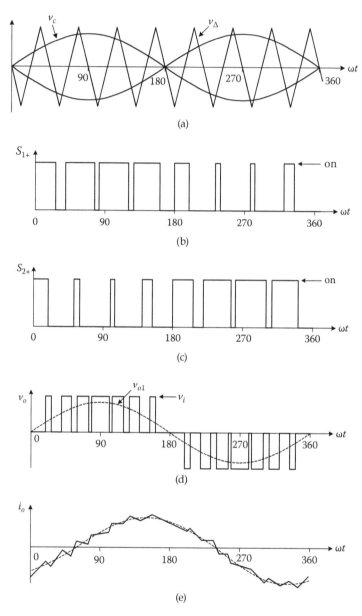

FIGURE 3.5
The full-bridge VSI ($m_a = 0.8$, $m_f = 8$).

The first pulse width to switch on and switch off S_1+ and S_1- is determined by the equation below:

$$0.8 sin100\pi t = 1600t-2 \tag{3.7}$$

The first pulse width to switch on and switch off the switches S_{2+} and S_{2-} is determined by the equation below:

$$-0.8 sin100\pi t = 1600t-2$$

or

$$0.8 sin100\pi t = 2-1600t \tag{3.8}$$

Since the output voltage varies between legs, the rms voltages of the output voltage harmonics are calculated by the following formula

$$(V_O)_h = \frac{2V_d}{\sqrt{2}} \frac{(\hat{V}_{AO})_h}{V_d/2} \tag{3.9}$$

where $(V_O)_h$ is the hth harmonic rms voltage of the output voltage, V_d is the DC link voltage and $(\hat{V}_{AO})_h/(V_d/2)$ is tabulated as a function of m_a, which can be found in Table 2.1.

Example 3.2

A single-phase full-bridge DC/AC inverter is shown in Figure 3.4 to implement SPWM with $V_d = 300$ V, $m_a = 1.0$ and $m_f = 31$. The fundamental frequency is 50 Hz. Determine the rms value of the fundamental frequency and some of the harmonics in output voltage using Table 3.1.

Solution:

From Equation 2.3 we have the general rms values

$$(V_O)_h = \frac{2V_d}{\sqrt{2}} \frac{(\hat{V}_{AO})_h}{V_d/2} = \frac{600}{\sqrt{2}} \frac{(\hat{V}_{AO})_h}{V_d/2} = 424.26 \frac{(\hat{V}_{AO})_h}{V_d/2} V$$

Checking the data from Table 2.1, we can get the rms values as follows:

TABLE 3.1

Iterative Method

t (ms/degree)	x	y	x|:y	Remarks
1.38889/25°	0.338	0.5	<	Decrease t
1.27778/23°	0.3126	0.3	>	Increase t
1.2889/23.2°	0.3152	0.32	<	Decrease t
1.2861/23.15°	0.3145	0.315	≈	

$(V_O)_1 = 424.26 \times 1.0 = 424.26V$ at 50 Hz

$(V_O)_{27} = 424.26 \times 0.018 = 7.64V$ at 1350 Hz

$(V_O)_{29} = 424.26 \times 0.318 = 134.92V$ at 1450 Hz

$(V_O)_{31} = 424.26 \times 0.601 = 254.98V$ at 1550 Hz

$(V_O)_{33} = 424.26 \times 0.318 = 134.92V$ at 1650 Hz

$(V_O)_{35} = 424.26 \times 0.018 = 7.64V$ at 1750 Hz

$(V_O)_{57} = 424.26 \times 0.033 = 14V$ at 2850 Hz

$(V_O)_{59} = 424.26 \times 0.212 = 89.94V$ at 2950 Hz

$(V_O)_{61} = 424.26 \times 0.181 = 76.79V$ at 3050 Hz

$(V_O)_{63} = 424.26 \times 0.181 = 76.79V$ at 3150 Hz

$(V_O)_{65} = 424.26 \times 0.212 = 89.94V$ at 3250 Hz

$(V_O)_{67} = 424.26 \times 0.033 = 14V$ at 3350 Hz,

etc.

3.2 Three-Phase Full-Bridge VSI

A three-phase full-bridge VSI is shown in Figure 3.6. The carrier-based pulse width modulation (PWM) technique is applied in this single-phase full-bridge VSI. Two large capacitors may be used to provide a neutral point N, but are not necessary. Six switches $S_1 - S_6$ in three legs are applied and switched by the PWM signal.

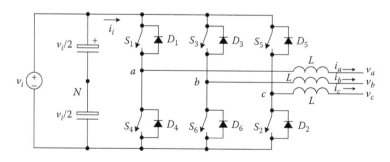

FIGURE 3.6

Three-phase full-bridge VSI. (a) Carrier and modulating signals. (b) Switch S_1/S_4 state. (c) Switch S_3/S_4 state. (d) AC output voltage. (e) AC output current.

Figure 3.7 shows the ideal waveforms associated with the full-bridge VSI. We can find out the phase delay between the output current and voltage.

Since the three-phase waveform in Figure 3.7a is not referring to the neutral point *N*, the operation conditions are different from single-phase half-bridge VSI. The maximum output line-to-line voltage is possibly greater than the half of the DC-link voltage. If it is operating in linear modulation, the output voltage is smaller than the DC-link voltage. The modulation indication

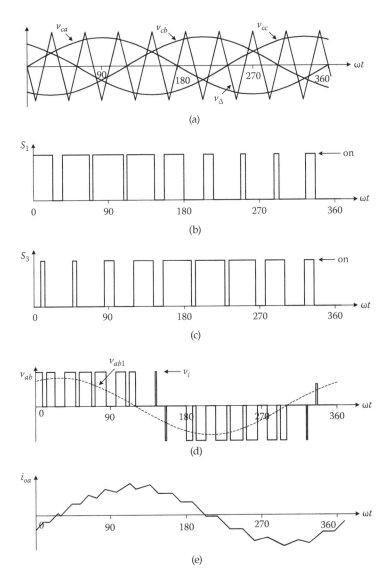

FIGURE 3.7
Three-phase full-bridge VSI ($m_a = 0.8$, $m_f = 9$).

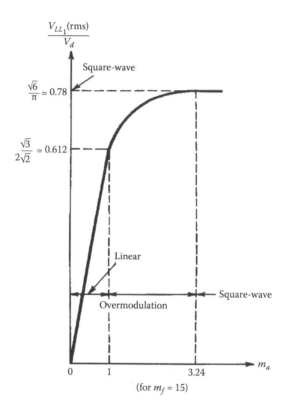

FIGURE 3.8
The function of modulation for a three-phase inverter.

of a three-phase VSI is different from that of single-phase half-bridge VSI as noted in Section 3.1.1. It is shown in Figure 3.8.

3.3 Vector Analysis and Determination of m_a

Vector analysis and determination of m_a are important topics for a three-phase PWM VSI. We discuss this topic in this section and would like to draw more attention to DC/AC inverter designs.

3.3.1 Vector Analysis

A three-phase PWM VSI vector is shown in Figure 3.9. Referring to Figure 3.6, we have

$$v_{control} = \hat{V}_{control} \sin(\omega_1 t) \tag{3.10}$$

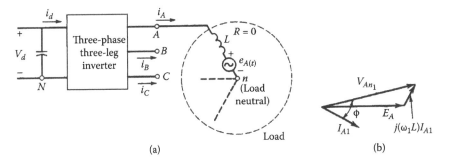

FIGURE 3.9
Three-phase inverter vector analysis: (a) circuit diagram; (b) phasor diagram (fundamental frequency).

Since

$$V_{AO} = \frac{v_{control}}{\hat{V}_{tri}} \frac{V_d}{2} = \frac{\hat{V}_{control}}{\hat{V}_{tri}} \sin(\omega_1 t) \frac{V_d}{2} = m_a \frac{V_d}{2} \sin(\omega_1 t) \qquad (3.11)$$

3.3.2 m_a Calculation

If the load is resistive, the phasor diagram is shown in Figure 3.10.

$$V_{An1} = m_a \frac{V_d}{2} \sin(\omega_1 t) \qquad (3.12)$$

$$V_{An1-rms} = m_a \frac{V_d}{2\sqrt{2}} \qquad (3.13)$$

$$V_{R-rms} = m_a \frac{V_d}{2\sqrt{2}} \cos\Phi \qquad (3.14)$$

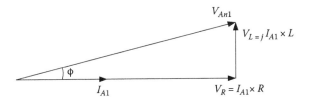

FIGURE 3.10
Three-phase inverter vector analysis for resistive load R.

$$\Phi V_{R-rms} = \tan^{-1}\left(\frac{\omega_1 L}{R}\right) = \tan^{-1}\left(\frac{100\pi * 20m}{80}\right) \approx 4.49° \tag{3.15}$$

$$V_{AB-rms} = \sqrt{3}V_{R-rms} = m_a \frac{\sqrt{3}V_d}{2\sqrt{2}}\cos\Phi \tag{3.16}$$

Hence,

$$m_a = \sqrt{3}V_{R-rms} = \frac{V_{AB-rms}}{V_d}\frac{2\sqrt{2}}{\sqrt{3}\cos\Phi} \approx 0.936 \tag{3.17}$$

We can choose $m_a = 0.939$.

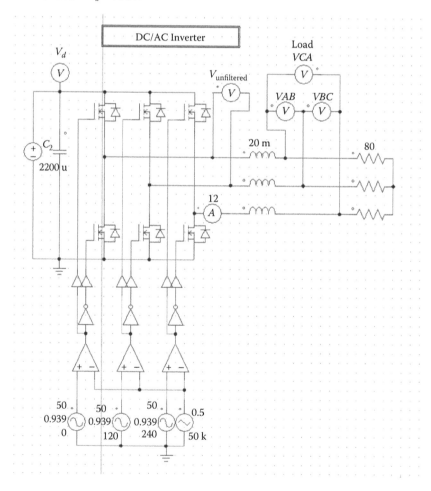

FIGURE 3.11
Three-phase inverter simulation circuit with a resistive load $R = 80\ \Omega$.

3.3.3 m_a Calculation with L-C Filter

If we set a L-C output filter (star connection) before the load, we have

$$R//C = \frac{R\frac{1}{j\omega C}}{R + \frac{1}{j\omega C}} = \frac{R}{1 + j\omega CR} \qquad (3.18)$$

$$V_R = \frac{R//C}{j\omega L + R//C} V_{An} = \frac{\frac{R}{1+j\omega CR}}{j\omega L + \frac{R}{1+j\omega CR}} V_{An} = \frac{R}{R(1 - \omega^2 CL) + j\omega L} V_{An} \qquad (3.19)$$

The variation is $\omega C \omega L = 100\pi \times 3u \times 100\pi \times 20m \approx 0.0056$.
Therefore, we can choose $m_a = 0.9304$.

3.3.4 Some Waveforms

We chose the DC input voltage $V_d = 700$ V, the load $R = 80$ Ω, and an L filter $L = 20$ mH. The simulation circuit is shown in Figure 3.11, and the output voltage is a three-phase 50 Hz/400 V (rms liner-to-liner) as shown in Figure 3.12.

In order to obtain better output voltage, we can set an L-C filter before the load. The capacitance is 3 μF in a star connection. The simulation circuit is

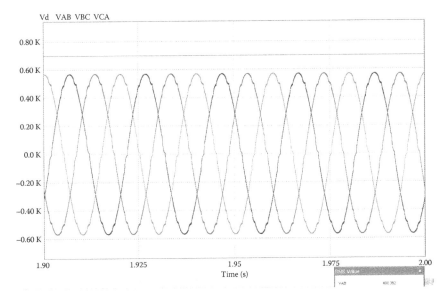

FIGURE 3.12
The corresponding input and output waveforms of VSI in Figure 3.11.

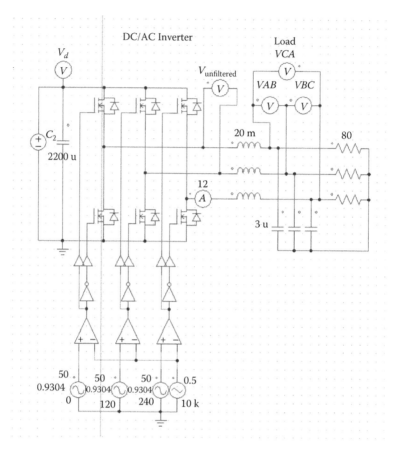

FIGURE 3.13
Three-phase inverter simulation circuit with a resistive load R = 80 Ω.

shown in Figure 3.13, and the output voltage is a three-phase 50 Hz/400 V (liner-to-liner rms) as shown in Figure 3.14.

We can also set an L-C filter before the load. The capacitance is 1 µF in delta connection. The simulation circuit is shown in Figure 3.15, and the output voltage is a three-phase 50 Hz/400 V (liner-to-liner rms) as shown in Figure 3.16.

3.4 Multistage PWM Inverter

Multistage PWM inverters can be constructed by two methods: multicell and multilevel. Unipolar modulation PWM inverters can be considered multistage inverters.

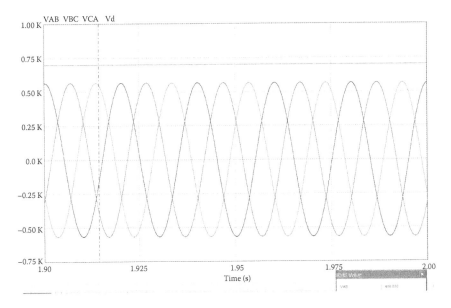

FIGURE 3.14
The corresponding input and output waveforms of VSI in Figure 3.13.

3.4.1 Unipolar PWM VSI

In Section 3.1 we introduced the single-phase source inverter operating in the bipolar modulation. Refer to the circuit in Figure 3.1; both upper switch S+ and lower switch S– work together. The carrier and modulating signals are shown in Figure 3.2a, and the switching signals for upper switch S+ and lower switch S– are shown in Figures 3.2b and c. The output voltage of the inverter is the pulse train with both polarities shown in Figure 3.2d.

There are some drawbacks using bipolar modulation: (1) if the inverter is VSI, a dead time has to be set to avoid a short circuit; (2) the zero output voltage corresponds to the equal-pulse width of positive and negative pulses; (3) power losses are high since two devices work, and hence the efficiency is lower; and (4) two devices should be controlled simultaneously.

The modulation method of the unipolar VSI is shown in Figure 3.17. There are two triangle signals in positive and negative levels. When the control signal is positive, only the top switch works and when the control signal is negative, only the lower switch works.

In most industrial applications, unipolar modulation is widely applied. The regulation and corresponding waveforms are shown in Figure 3.17 with $m_a = 0.8$ and $m_f = 9$. For unipolar regulation the m_a is measured by

$$m_a = \frac{V_{in-m}}{2V_{tri-m}} \tag{3.20}$$

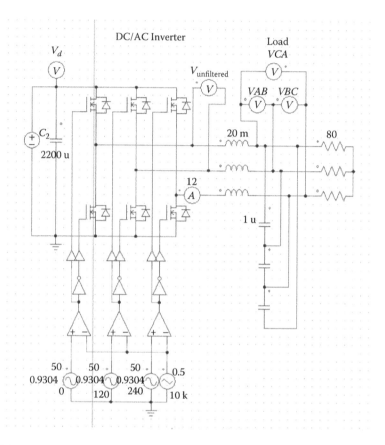

FIGURE 3.15
Three-phase inverter simulation circuit with a resistive load $R = 80 \ \Omega$.

This regulation method is likely a two-stage PWM inverter. If the output voltage is positive, only the upper device works and the lower device idles. Therefore, the output voltage only remains the positive polarity pulse train. Conversely, if the output voltage is negative, only the lower device works and the upper device idles. Therefore, the output voltage only remains the negative polarity pulse train. The advantages to implement unipolar regulation are

- No need to set dead time.
- The pulses are narrow, for example, zero output voltage requires zero pulse width.
- Power losses are low and hence the efficiency is high.
- Only one device should be controlled in half-cycle operation.

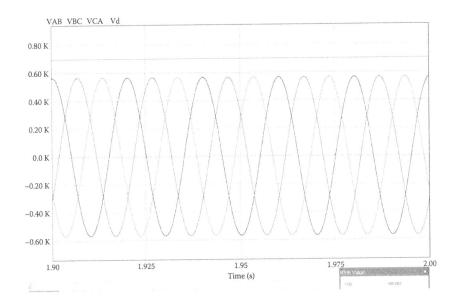

FIGURE 3.16
The corresponding input and output waveforms of VSI in Figure 3.15.

3.4.2 Multicell PWM VSI

Multistage PWM inverters can consist of many cells. Each cell can be a single-phase or three-phase input plus single-phase output VSI, which is shown in Figure 3.18. If the three-phase AC supply is a secondary winding of a main transformer, it is floating and isolated from other cells and a common ground point. Therefore, all cells can be linked in a series or parallel manner.

A three-stage PWM inverter is shown in Figure 3.19. Each phase consists of three cells with difference phase-angle shift by 20° of each other.

The carrier-based pulse width modulation (PWM) technique is applied in this three-phase multistage PWM inverter. Figure 3.20 shows the ideal waveforms associated with the full-bridge VSI. We can find out the output of the phase delayed between the output current and voltage.

3.4.3 Multilevel PWM Inverter

A three-level PWM inverter is shown in Figure 3.21. The carrier-based pulse width modulation (PWM) technique is applied in this multilevel PWM inverter. Figure 3.22 shows the ideal waveforms associated with the multilevel PWM inverter. We can determine the output of the phase delayed between the output current and voltage.

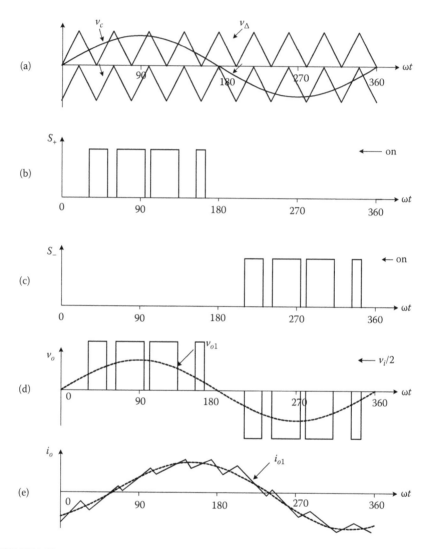

FIGURE 3.17
Three-phase unipolar regulation inverter ($m_a = 0.8$, $m_f = 9$).

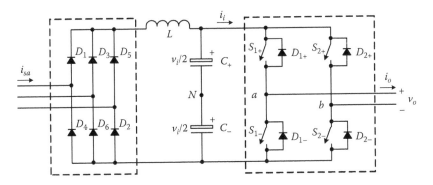

FIGURE 3.18
Three-phase input single-phase output cell.

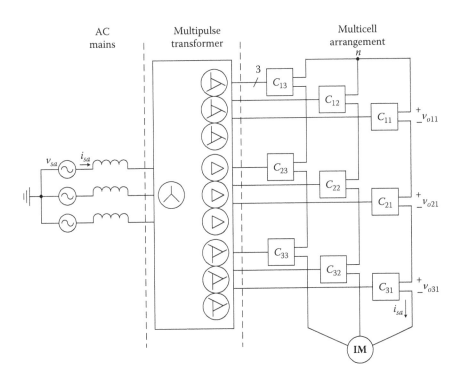

FIGURE 3.19
Multistage converter based on a multicell arrangement.

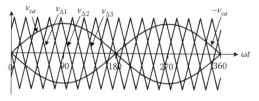

(a) Carrier and modulating signals

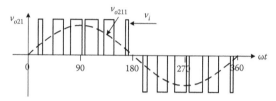

(b) Cell c_{11} AC output voltage

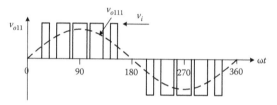

(c) Cell c_{21} AC output voltage

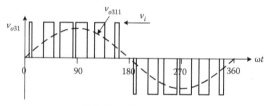

(d) Cell c_{31} AC output voltage

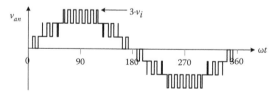

(e) Phase a load voltage

FIGURE 3.20
Multicell PWM inverter (3 stages, $m_a = 0.8$, $m_f = 6$).

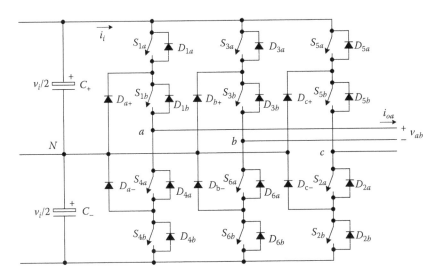

FIGURE 3.21
Three-phase, three-level VSI. (a) Carrier and modulating signals. (b) Switch S_{1a} status. (c) Switch S_{4b} status. (d) Inverter phase a-N voltage. (e) AC output line voltage. (f) AC output phase voltage.

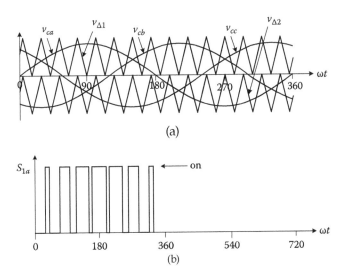

FIGURE 3.22
Three-level VSI (3 levels, $m_a = 0.8$, $m_f = 15$).

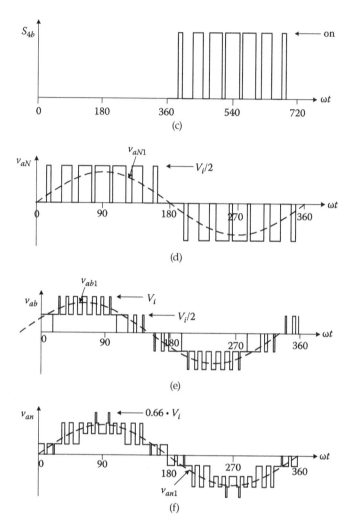

FIGURE 3.22 (*continued*)
Three-level VSI (3 levels, $m_a = 0.8$, $m_f = 15$).

References

1. Mohan, N., Undeland, T. M., and Robbins, W. P. 2003. *Power Electronics: Converters, Applications and Design (3rd edition)*. New York: John Wiley & Sons.
2. Holtz, J. 1992. Pulsewidth modulation—a survey. *IEEE Trans. Ind. Electron.*, pp. 410–420.
3. Luo, F. L. and Ye, H. 2010. *Power Electronics: Advanced Conversion Technologies*. Boca Raton, FL: Taylor & Francis.

4

Current Source Inverters

Current source inverters (CSIs) have the boost function, but they are not widely applied in industrial applications. Therefore, they are not well discussed in the literature. We have carried out an investigation and obtained the output voltage, which is higher than the input voltage.

4.1 Three-Phase Full-Bridge Current Source Inverter

A typical three-phase full-bridge current source inverter is shown in Figure 4.1 [1,2].

The carrier-based pulse width modulation (PWM) technique is applied in this three-phase full-bridge CSI. The main objective of these static power converters is to produce AC output current waveforms from a DC current power supply. Six switches $S_1 - S_6$ are applied and switched by the PWM signal. Figure 4.2 shows the ideal waveforms associated with the full-bridge CSI [1,2].

The CSI has a boost function. Usually, the output voltage can be higher than the input voltage. We can find out the phase ahead between the output voltage and current.

4.2 Boost-Type CSI

There are two methods introduced in this section: negative polarity input voltage and positive polarity input voltage.

4.2.1 Negative Polarity Input Voltage

Figure 4.3 shows the circuit of the boost-type CSI with negative polarity input voltage [3]. The circuit (v_i, L, C, and D plus the load) is a DC/DC boost converter. We point out that the polarity of the input voltage v_i is negative. The real voltage v_d applied to the following PWM main circuit can be higher than the input voltage v_i. Actually, the topology from v_d to the output voltage (v_a, v_b, and v_c) is a VSI.

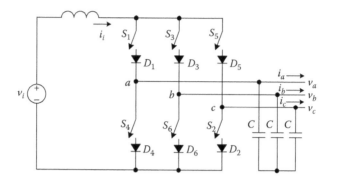

FIGURE 4.1
Three-phase CSI.

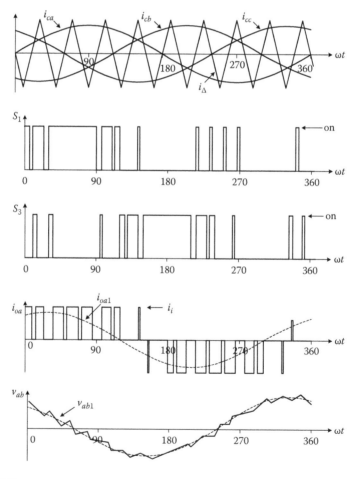

FIGURE 4.2
Three-phase CSI ($m_a = 0.8$, $m_f = 9$).

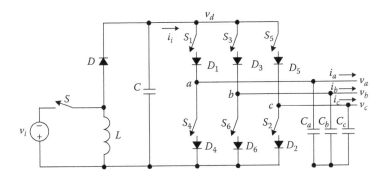

FIGURE 4.3
Boost-type CSI with negative input voltage.

For example, we set $k = 0.67$ and obtain $V_d = 400$ V, and the output voltage is three-phase 50 Hz/246.6 V (rms, line-to-line). The simulation circuit is shown in Figure 4.4 and waveforms of the input voltage V_{in}, the middle voltage V_d and output voltages are shown in Figure 4.5.

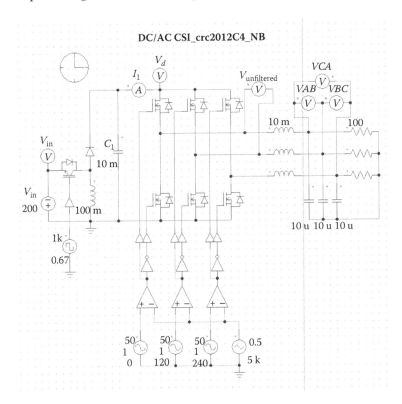

FIGURE 4.4
Simulation circuit of a boost-type CSI with negative input voltage.

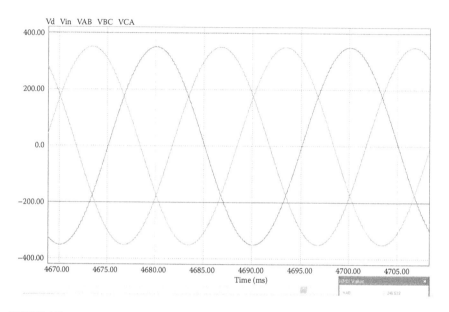

FIGURE 4.5
Waveforms of a boost-type CSI with negative input voltage.

4.2.2 Positive Polarity Input Voltage

Figure 4.6 shows the circuit of the boost-type CSI with positive polarity input voltage. The circuit (v_i, L, C, and D plus the load) is a DC/DC boost converter. We pinpoint that the polarity of the input voltage v_i is positive. The real voltage v_d applied to the following PWM main circuit can be higher than the input voltage v_i. Actually, the topology from v_d to the output voltage (v_a, v_b, and v_c) is a VSI.

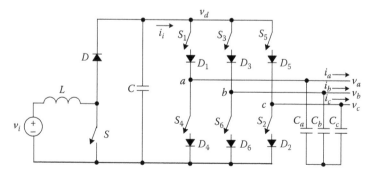

FIGURE 4.6
Boost-type CSI with positive input voltage.

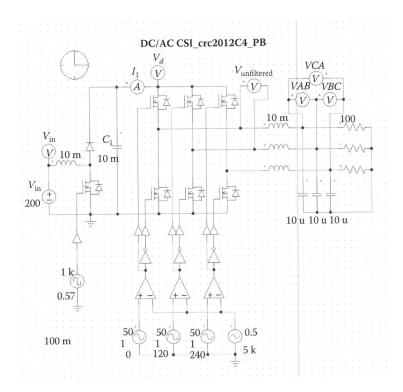

FIGURE 4.7
Simulation circuit of a boost-type CSI with positive input voltage.

For example, we sent $k = 0.57$ and obtain $V_d = 475$ V, and the output voltage is three-phase 50 Hz/294 V (rms, line-to-line). The simulation circuit is shown in Figure 4.7, and the waveforms of the input voltage V_{in}, the middle voltage V_d, and output voltages are shown in Figure 4.8.

4.3 CSI with L-C Filter

The third type circuit is CSI with L-C filter as shown in Figure 4.9, which is derived from Figure 4.1 [4,5]. Applying the L-C filter the output voltage can be higher than the input voltage. The simulation circuit is shown in Figure 4.10, and the waveforms of the input voltage V_{dc1}, the middle voltage V_{dc2} and output voltages are shown in Figure 4.11.

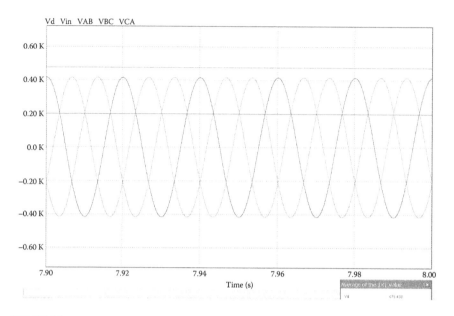

FIGURE 4.8
Waveforms of a boost-type CSI with positive input voltage.

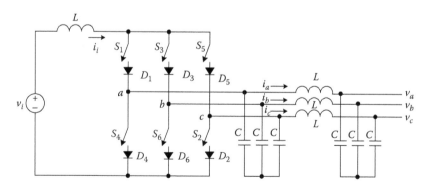

FIGURE 4.9
CSI with L-C filter.

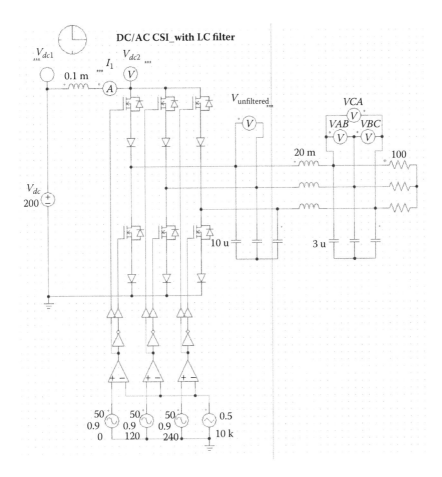

FIGURE 4.10
Simulation circuit of a CSI with L-C filter.

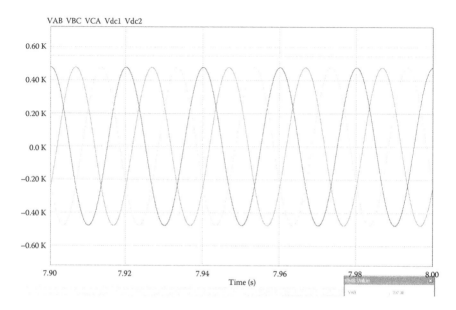

FIGURE 4.11
Waveforms of a CSI with L-C filter.

References

1. Mohan, N., Undeland, T. M., and Robbins, W. P. 2003. *Power Electronics: Converters, Applications and Design (3rd edition)*. New York: John Wiley & Sons.
2. Luo, F. L. and Ye, H. 2010. *Power Electronics: Advanced Conversion Technologies*. Boca Raton, FL: Taylor & Francis.
3. Yu, R., Wei, X., and Wu, X. 2009. *Variable Frequency Current Source Inverter for Grid-Connected PV System*. ICEET, pp. 267–370.
4. Mirafzal B., Saghaleini M. and Kaviani A. K. 2011. An SVPWM-based switching pattern for stand-alone and grid-connected three-phase single-stage boost inverters. *IEEE Trans. Power Electron.*, pp. 1101–1111.
5. Saghaleini, M. and Mirafzal, B. 2011. Power control in three-phase grid-connected current-source boost inverter. *IEEE Conference*, pp. 776–783.

5

Impedance Source Inverters

Impedance source inverter (ZSI) is a new approach to DC/AC conversion technology. It was published by Peng in 2003 [1–3]. The ZSI circuit diagram shown in Figure 5.1 consists of an x-shaped impedance network formed by two capacitors and two inductors, and it provides unique buck–boost characteristics. Moreover, unlike VSI, the need for dead time would not arise with this topology. Due to these attractive features, it has found numerous industrial applications including variable speed drives and distributed generation (DG). However, it has not been widely researched as a DG topology. Moreover, all these industrial applications require proper closed loop controlling to adjust their operating conditions subject to changes in both input and output conditions. On the other hand, the presence of the x-shaped impedance network and the need for short-circuiting the inverter arm to boost the voltage would complicate the controlling of ZSI.

5.1 Comparison with VSI and CSI

ZSI is a new inverter different from traditional voltage source inverter (VSI) and current source inverter (CSI) devices. In order to understand ZSI's advantages, it is necessary to compare it with VSI and CSI [3,4].

The VSI has the following conceptual and theoretical barriers and limitations:

1. The AC output voltage cannot exceed the DC-link voltage. Therefore, the VSI is a buck (step-down) inverter for DC/AC power conversion. For applications where overdrive is desirable and the available DC voltage is limited, an additional DC/DC boost converter is needed to obtain a desired AC output. The additional power converter stage increases system cost and lowers efficiency.

2. The upper and lower devices of each phase leg cannot be gated on simultaneously either by purpose or by EMI noise. Otherwise, a shoot-through would occur and destroy the devices. The shoot-through

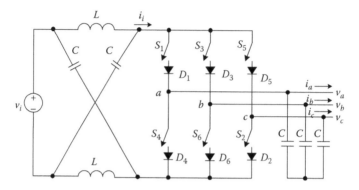

FIGURE 5.1
Impedance source inverter (ZSI).

problem due to electromagnetic interference (EMI) noise misgating on is a major killer to the converter's reliability. Dead time to block both upper and lower devices has to be provided in the VSI, which causes waveform distortion, etc.

3. An output L-C filter is needed to provide a sinusoidal voltage compared with the current source inverter, which causes additional power loss and control complexity.

The CSI has the following conceptual and theoretical barriers and limitations:

1. The AC output voltage has to be greater than the original DC voltage that feeds the DC inductor, or the DC voltage produced will always be always smaller than the AC input voltage. Therefore, the CSI is a boost inverter for DC/AC power conversion. For applications where a wide voltage range is desirable, an additional DC–DC buck (or boost) converter is needed. The additional power conversion stage increases system cost and lowers efficiency.

2. At least one of the upper devices and one of the lower devices must be gated on and maintained on at any time. Otherwise, an open circuit of the DC inductor would occur and destroy the devices. The open circuit problem by EMI noise misgating off is a major concern from the view of the converter's reliability. Overlap time for safe current commutation is needed in the I-source converter, which also causes waveform distortion, etc.

3. The main switches of the I-source converter have to block reverse voltage that requires a series diode to be used in combination with high-speed and high-performance transistors such as insulated gate bipolar transistors (IGBTs). This avoids the direct use of low-cost and high-performance performance IGBT modules and intelligent power modules (IPMs).

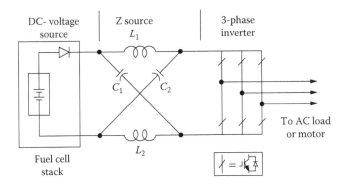

FIGURE 5.2
Z-source inverter for fuel cell applications.

In addition, both the VSI and the CSI have the following common problems:

1. They work as either a boost or a buck converter and cannot work as a buck-and-boost converter. That is, their obtainable output voltage range must be either greater or smaller than the input voltage.
2. Their main circuits cannot be used interchangeably. In other words, the VSI main circuit cannot be used for the CSI, and vice versa.
3. They are vulnerable to EMI noise that affects reliability.

To overcome the above problems of the traditional VSI and CSI, ZSI was designed as shown in Figure 5.2. It employs a unique impedance network to couple the converter main circuit to the power source. The ZSI overcomes the above-mentioned conceptual and theoretical barriers and limitations of the traditional VSI and CSI, and provides a novel power conversion concept.

In Figure 5.2, a two-port network that consists of split inductors L_1 and L_2 and capacitors C_1 and C_2 connected in an x shape is employed to provide an impedance source (Z-source) coupling the converter (or inverter) to the DC source. Switches used in ZSI can be a combination of switching devices and diodes. Actually, if the two inductors have zero inductance, the ZSI becomes a VSI and if the two capacitors have zero capacitance, the ZSI becomes a CSI. The advantages of the ZSI are listed as follows:

1. The AC output voltage is not fixed lower or higher than the DC-link (or DC source) voltage. Therefore, the ZSI is a buck–boost inverter for DC/AC power conversion. For applications where overdrive is desirable and the available DC voltage is not limited, there is no need for an additional DC/DC boost converter to obtain a desired AC output. Therefore, the system cost is low and efficiency is high.

2. The Z-circuit consist of two inductors and two capacitors and can restrict the overvoltage and overcurrent. Therefore, the legs in the main bridge can operate in short circuit and open circuit in a short time. There is no restriction for the main bridge such as dead time for VSI and overlap time for CSI.

3. ZSI has an anti-noise function. The shoot-through problem by electromagnetic interference (EMI) noise misgating on will not damage the device or the converter's reliability.

5.2 Equivalent Circuit and Operation

A three-phase ZSI used for fuel cell application is shown in Figure 5.3. It has 9 permissible switching states (vectors): 6 active vectors as a traditional VSI has plus 3 zero vectors when the load terminals are shorted through both the upper and lower devices of any one-phase leg (i.e., both devices are gated on), any two-phase legs, or all three-phase legs. This shoot-through zero state (or vector) is undesirable in the traditional VSI, because it would cause a shoot-through. We call this third zero state (vector) the shoot-through zero state (or vector), which can be generated seven different ways: shoot-through via any one-phase leg, combinations of any two-phase legs, and all three-phase legs. The Z-source network makes the shoot-through zero state possible. This shoot-through zero state provides the unique buck–boost feature to the inverter.

Figure 5.3 shows the equivalent circuit of the ZSI shown in Figure 5.2 when viewed from the DC link. The inverter bridge is equivalent to a short

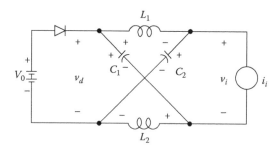

FIGURE 5.3
Equivalent circuit of the Z-source inverter viewed from the DC link.

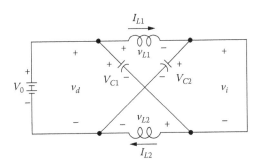

FIGURE 5.4
Equivalent circuit of the Z-source inverter viewed from the DC link when the inverter bridge is in the shoot-through zero state.

circuit when the inverter bridge is in the shoot-through zero state, as shown in Figure 5.4, whereas the inverter bridge becomes an equivalent current source as shown in Figure 5.5 when in one of the six active states. Note that the inverter bridge can be also represented by a current source with zero value (i.e., an open circuit) when it is in one of the two traditional zero states. Therefore, Figure 5.5 shows the equivalent circuit of the Z-source inverter viewed from the DC link when the inverter bridge is in one of the eight non-shoot-through switching states.

All the traditional pulse width modulation (PWM) schemes can be used to control the Z-source inverter, and their theoretical input–output relationships still hold. Figure 5.6 shows the traditional PWM switching sequence based on the triangular carrier method. In every switching cycle, the two

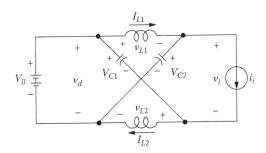

FIGURE 5.5
Equivalent circuit of the Z-source inverter viewed from the DC link when the inverter bridge is in one of the eight non-shoot-through switching states.

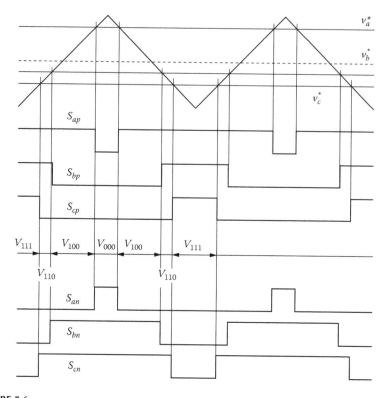

FIGURE 5.6
Traditional carrier-based PWM control without shoot-through zero states, where the traditional zero states (vectors) V111 and V000 are generated every switching cycle and determined by the references.

non-shoot-through zero states are used along with two adjacent active states to generate the desired voltage. When the DC voltage is high enough to generate the desired AC voltage, the traditional PWM of Figure 5.6 is used. While the DC voltage is not enough to directly generate a desired output voltage, a modified PWM with shoot-through zero states will be used as shown in Figure 5.7 to boost voltage. It should be noted that each phase leg still switches on and off once per switching cycle. Without changing the total zero-state time interval, shoot-through zero states are evenly allocated into each phase. That is, the active states are unchanged. However, the equivalent DC-link voltage to the inverter is boosted because of the shoot-through states. The detailed relationship will be analyzed in the next section. It is noticeable here that the equivalent switching frequency viewed from the Z-source network is six times the switching frequency of the main inverter, which greatly reduces the required inductance of the Z-source network.

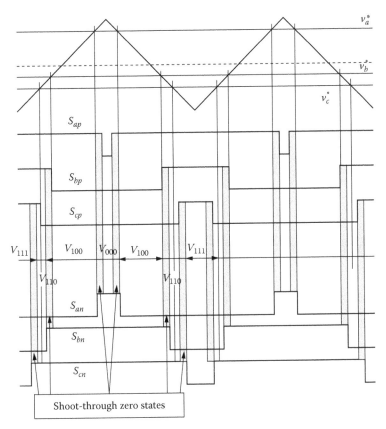

FIGURE 5.7
Modified carrier-based PWM control with shoot-through zero states that are evenly distributed among the three phase legs, while the equivalent active vectors are unchanged.

5.3 Circuit Analysis and Calculations

Assuming that the inductors L_1 and L_2 and capacitors C_1 and C_2 have the same
inductance L and capacitance C, respectively, the Z-source network becomes
symmetrical. From the symmetry and the equivalent circuits, we have

$$V_{C1} = V_{C2} = V_C \quad v_{L1} = v_{L2} = v_L \tag{5.1}$$

Assume that the inverter bridge is in the shoot-through *zero state* for an interval of T_0 during a switching cycle T. From the equivalent circuit in Figure 5.4,
one has

$$v_L = V_C \quad v_d = 2V_C \quad v_i = 0 \tag{5.2}$$

Now consider that the inverter bridge is in one of the eight non-shoot-through states for an interval of T_1 during the switching cycle T. From the equivalent circuit in Figure 5.4, one has

$$v_L = V_0 - V_C \quad v_d = V_0 \quad v_i = V_C - v_L = 2V_C - V_0 \tag{5.3}$$

where V_0 is the DC source voltage and $T = T_0 + T_1$. The switching duty cycle $k = T_1/T$.

The average voltage of the inductors over one switching period should be zero in the steady state, from Equations (5.2) and (5.3); thus, we have

$$V_L = \bar{v}_L = \frac{T_0 V_C + T_1(V_0 - V_C)}{T} = 0 \tag{5.4}$$

or

$$\frac{V_C}{V_0} = \frac{T_1}{T_1 - T_0} \tag{5.5}$$

Similarly, the average DC-link voltage across the inverter bridge can be found as follows:

$$V_i = \bar{v}_i = \frac{T_0 \times 0 + T_1(2V_C - V_0)}{T} = \frac{T_1}{T_1 - T_0} V_0 = V_C \tag{5.6}$$

The peak DC-link voltage across the inverter bridge is expressed in Equation (5.3) and can be rewritten as

$$\hat{v}_i = V_C - v_L = 2V_C - V_0 = \frac{T}{T_1 - T_0} V_0 = B \cdot V_0 \tag{5.7}$$

where

$$B = \frac{T}{T_1 - T_0} = \frac{1}{1 - 2\frac{T_0}{T}} \geq 1 \tag{5.8}$$

B is the boost factor resulting from the shoot-through zero state. Usually, T_1 is greater than T_0, that is, $T_0 < T/2$. The peak DC-link voltage $\hat{v}_i$ is the equivalent DC-link voltage of the inverter. On the other side, the output peak phase voltage from the inverter can be expressed as

$$\hat{v}_{ac} = M \cdot \frac{\hat{v}_i}{2} \tag{5.9}$$

where M is the modulation index. Using Equation (5.7), Equation (5.9) can be further expressed as

$$\hat{v}_{ac} = M \cdot B \cdot \frac{V_0}{2} \tag{5.10}$$

For the traditional VSI, we have the well-known relationship: $\hat{v}_{ac} = M \cdot \frac{V_0}{2}$. Equation (5.10) shows that the output voltage can be stepped up and down by choosing an appropriate buck–boost factor MB.

$$MB = \frac{T}{T_1 - T_0} M \tag{5.11}$$

MB is changeable from 0 to ∞. From Equations (5.1), (5.5), and (5.8), the capacitor voltage can be expressed as

$$V_C = \frac{1 - \frac{T_1}{T}}{1 - 2\frac{T_0}{T}} V_0 \tag{5.12}$$

The buck–boost factor MB is determined by the modulation index M and boost factor B. The boost factor B as expressed in Equation (5.8) can be controlled by the duty cycle (i.e., interval ratio) of the shoot-through zero state over the non-shoot-through states of the inverter PWM.

Note that the shoot-through zero state does not affect the PWM control of the inverter, because it equivalently produces the same zero voltage to the load terminal. The available shoot-through period is limited by the zero-state period that is determined by the modulation index.

5.4 Simulation and Experimental Results

Simulations have been performed to confirm the above analysis. Figure 5.8 shows the circuit configuration, and Figure 5.9 shows simulation waveforms when the fuel cell stack voltage is $V_0 = 150$ V and the Z-source network parameters are $L_1 = L_2 = L = 160$ µH and $C_1 = C_2 = C = 1000$ µF.

The purpose of the system is to produce a three-phase, 208 V rms power from the fuel cell stack whose voltage changes from 150 to 340 V DC depending on load current. From the simulation waveforms of Figure 5.9, it is clear that the capacitor voltage was boosted to $V_{C2} = 335$ V and the output line-to-line was 208 V rms or 294 V peak. In this case, the modulation index was set to $M = 0.642$, and the shoot-through duty cycle was set to $T_0/T = 0.358$, and

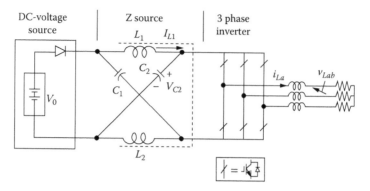

FIGURE 5.8
Simulation and prototype system configuration.

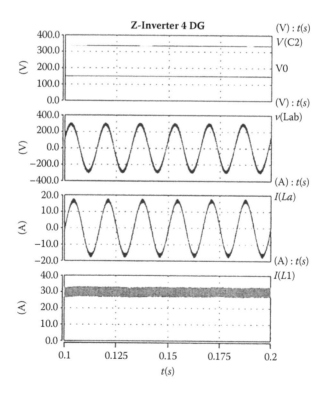

FIGURE 5.9
Simulation waveforms when the fuel cell voltage $V = 150$ V, inverter modulation index $M = 0.642$, and shoot-through duty cycle $T_0/T = 0.358$.

switching frequency was 10 kHz. The shoot-through zero state was populated evenly among the three phase legs, achieving an equivalent switching frequency of 60 kHz viewed from the Z-source network. Therefore, the required DC inductance is minimized. From the above analysis, we have the following theoretical calculations:

$$B = \frac{1}{1-2\frac{T_0}{T}} = \frac{1}{1-0.716} = 3.55 \tag{5.13}$$

$$V_{C1} = V_{C1} = V_C \frac{1-\frac{T_0}{T}}{1-2\frac{T_0}{T}} V_0 = \frac{1-0.358}{1-0.716} 150 = 339 \text{ V} \tag{5.14}$$

$$\hat{V}_{ac} = MB\frac{V_0}{2} = 0.642 \times 0.358\frac{150}{2} = 169.5 \text{ V} \tag{5.15}$$

Equation (5.15) is the phase peak voltage, which implies that the line-to-line voltage is 208 V rms or 294 V peak. The above theoretical values are quite consistent with the simulation results. The simulation proved the ZSI concept.

A prototype as shown in Figure 5.8 has been constructed. The same parameters as the simulation were used. Figures 5.10 and 5.11 show experimental results. When the fuel cell voltage is low, as shown in Figure 5.10, the

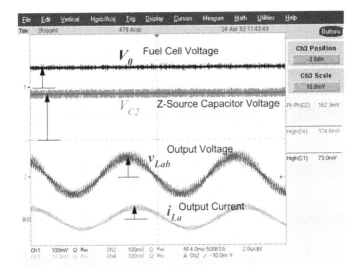

FIGURE 5.10
Experimental waveforms when the fuel cell voltage is low, inverter modulation index $M =$ 0.642, and shoot-through period ratio $T_0/T = 0.358$ (V_0 and V_{C2}: 200V/div, V_{Lab}: 2* 200V/div, i_{La}: 50 A/div, and time: 4 ms/div).

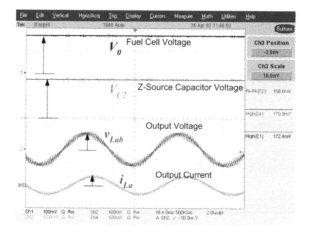

FIGURE 5.11

Experimental waveforms when the fuel cell voltage is high. Inverter modulation index $M = 1.0$ without using the shoot-through state or shoot-through period ratio $T_0/T = 0$.

shoot-through state was used to boost the voltage in order to maintain the desired output voltage. The waveforms are consistent with the simulation results. When the fuel-cell voltage is high enough to produce the desired output voltage, the shoot-through state was not used, as shown in Figure 5.11, where the traditional PWM control without shoot-through was used. By controlling the shoot-through state duty cycle or the boost factor, the desired output voltage can be obtained regardless of the fuel cell voltage.

References

1. Peng, F. Z. 2003. Z-source inverter. *IEEE Trans. Ind. Applicat.*, pp. 504–510.
2. Trzynadlowski, A. M. 1998. *Introduction to Modern Power Electronics*. New York: John Wiley & Sons.
3. Anderson, J. and Peng, F. Z. 2008. Four quasi-Z-source inverters. *Proc. IEEE PESC'2008*, pp. 2743–2749.
4. Luo, F. L. and Ye, H. 2010. *Power Electronics: Advanced Conversion Technologies*. Boca Raton, FL: Taylor & Francis.

6

Quasi-Impedance Source Inverters

In recent years, many researchers have worked in many directions to develop impedance source inverters (ZSIs) to achieve different objectives [1–8]. Some have worked on developing different kinds of topological variations, whereas others have worked on developing ZSI into different applications where controller design, modeling and analyzing its operating modes, and developing modulation methods are addressed. Theoretically, ZSI can produce infinite gain like many other DC/DC boosting topologies; however, it is well known that this cannot be achieved due to effects of parasitic components where the gain tends to drop drastically [1]. Conversely, high boost could increase power losses and instability. On the other hand, shoot-through interval, the variable that is responsible for increasing the gain and is interdependent with the other variable modulation index that controls the output of the ZSI, also imposes limitations on variability and thereby the boosting of output voltage. That is, an increase in boosting factor would compromise the modulation index and result in lower modulation index [2]. Also, the voltage stress on the switches would be high due to the pulsating nature of the output voltage.

Unlike the DC/DC converters, so far researchers of ZSIs have not concentrated on improving the gain of the converter. This opens a significant research gap in the field of ZSI development, especially in some applications such as solar and fuel cells where generated power is integrated into the grid and may require high voltage gain to match the voltage difference and also to compensate for the voltage variations. Its effect is significant when such sources are connected to 415 V three-phase systems. In the case of fuel and solar cells, although it is possible to increase the number of cells to increase the voltage, there are other influencing factors that need to be taken into account. Sometimes, the available number of cells is limited, or environmental factors could come into play due to shading of light for some cells, which could result in poor overall energy catchment. Some manufacturers produce fuel cells with lower voltage to achieve a faster response. Such factors could demand power converters with a larger boosting ratio. This cannot be realized with a single ZSI. Hence, this chapter focuses on developing a new family of ZSIs that would realize extended boosting capability.

6.1 Introduction to ZSI and Basic Topologies

The basic topology of ZSI was originally proposed by Peng. This is a single-stage buck–boost topology due to the presence of the x-shaped impedance network as shown in Figure 6.1a, which allows the safe shoot-through of inverter arms, avoiding the dead time that was needed in traditional VSI. However, unlike the VSI, the original ZSI does not share the ground point of the DC source with the converter, and also the current drawn from the source will be discontinuous; these would be disadvantages in some applications, and it may be required to have a decoupling capacitor bank at the front end to avoid current discontinuity. Subsequently, the ZSI was modified as shown in Figure 6.1b and 6.1c, where now an impedance network is placed at the bottom or top arm of the inverter. The advantage of this topology is that in one topology ground point can be shared, and in both cases the voltage stress on the component is much lower compared to that of the traditional ZSI. However, the current discontinuity was still present, and so an alternative continuous current *quasi*-ZSI (qZSI) is proposed, but this continuous current circuit is not considered in developing new converters. In terms of topology, the qZSI has no disadvantage over the traditional topology. In this chapter, a discontinuous current qZSI inverter is used to extend the boosting capability. In summary, the proposed qZSIs operate similarly to the original ZSI, and the same modulation schemes can be applied.

6.2 Extended Boost qZSI Topologies

In this chapter, four new converter topologies are proposed. Mainly these topologies can be categorized into diode-assisted boost or capacitor-assisted boost, and then they can be further divided into continuous current and discontinuous current topologies. Their operation is extensively described in the coming sections. All these topologies can be modulated using the modulation methods proposed for the original ZSI. In this context, the modulation method proposed is used. The other advantage of the proposed new topologies is their expandability. This was not possible with the original ZSI; that is, if one needs additional boosting, another stage can be cascaded at the front end. The new topology would operate with the same number of active switches. The only addition would be one inductor, one capacitor, and two diodes for the diode-assisted case and one inductor, two capacitors, and one diode for the capacitor-assisted case for each added new stage. By defining the shoot-through duty ratio (D_S), then, for each added new stage, the boosting factor can be increased with a factor of $1/(1-D_S)$ in the case of the diode-assisted topology. Then the capacitor-assisted topology would have a

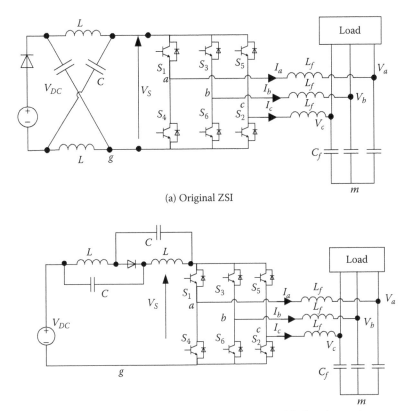

(a) Original ZSI

(b) Discontinuous current quasi Z-source inverter with shared ground

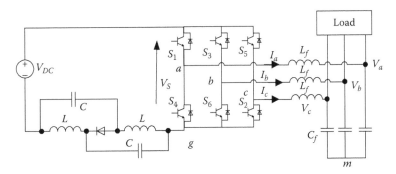

(c) Discontinuous current quasi ZSI with low voltage level at components

FIGURE 6.1
Various ZSIs.

boosting factor of $1/(1-3D_S)$ compared to $1/(1-2D_S)$ in the traditional topology. However, similar to the other boosting topologies, it is not advisable to operate with very high or very low shoot-through values. Also, careful consideration is needed in selecting the boosting factor modulation index for the suitable topology to achieve the high efficiency. These aspects need further research, and they will be addressed in a future paper.

6.2.1 Diode-Assisted Extended Boost qZSI Topologies

In this category, two new families of topologies are proposed, namely, the continuous current and the discontinuous current type topology. Figure 6.2 shows the continuous current type topology, and it can be extended to have very high boost by cascading more stages as shown in Figure 6.3. This new topology comprises an additional inductor, a capacitor, and two diodes. The operating principle of this additional impedance networks is similar to that found in cascaded boost and Luo converters [4–7]. The added impedance network provides the boosting function without disturbing the operation inverter.

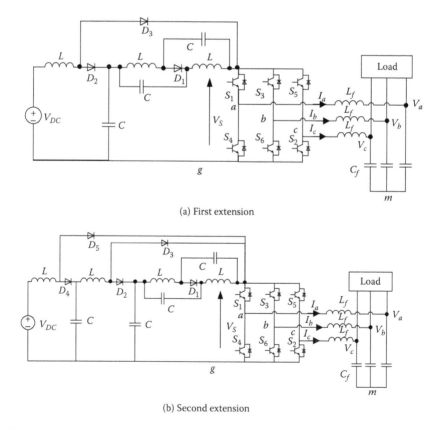

(a) First extension

(b) Second extension

FIGURE 6.2
Diode-assisted extended boost continuous current qZSI.

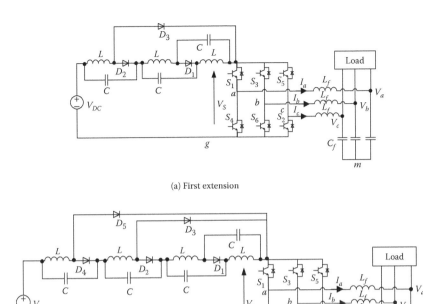

(a) First extension

(b) Second extension

FIGURE 6.3
Diode-assisted extended boost discontinuous current qZSI.

First, considering the continuous current topology and its steady-stage operation, we know that this converter has three operating states similar to those of the traditional ZSI topology. For simplicity, it can be classified into shoot-through and non-shoot-through states. Then the inverter's action is replaced by a current source plus a single switch. First, consider the non-shoot-through state, which is represented with an open switch. Also, diodes D_1 and D_2 are conducting, and D_3 is in blocking state; therefore, the inductors discharge, and the capacitors are charged. Figure 6.4b shows the equivalent circuit diagram for the non-shoot-through state.

By applying KVL, the following steady-state relationships can be observed: $V_{DC} + v_{L3} = V_{c3}$, $v_{L1} = V_{c1}$, $V_{L2} = V_{c2}$, and $V_S = V_{c3} + V_{c2} + V_{L1}$. Figure 6.4c shows the equivalent circuit diagram for the shoot-through state, where it is represented with the closed switch and D_3 is conducting, and D_1 and D_2 diodes are in blocking state where all the inductors get charged. Energy is transferred from the source to the inductor or the capacitor to the inductor, while capacitors are discharged. Similar relationships can be derived as $V_{DC} + v_{L3} = 0$, $V_{c3} + V_{L2} + V_{c1} = 0$, $V_{c3} + V_{c2} + V_{L1} = 0$, and $V_S = 0$, $V_{c3} + V_{c2} = V_{L1}$. Since the average voltage

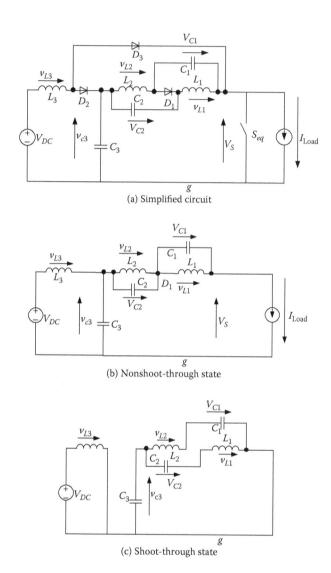

FIGURE 6.4
Simplified diagram of diode-assisted extended boost continuous current qZSI: (a) continuous current, (b) discontinuous current, (c) high-extended continuous current.

across the inductors is zero and, by defining the shoot-through duty ratio as D_S and non-shoot-through duty ratio as D_A where $D_A + D_S = 1$, the following relations can be derived:

$$V_{C3} = \frac{1}{1-D_S} V_{DC} \text{ and } V_{C1} = V_{C2} = \frac{D_S}{1-2D_S} V_{C3} = \frac{D_S}{(1-2D_S)(1-D_S)} V_{DC} \quad (6.1)$$

From the above equations, the peak voltage across the inverter $\hat{v}_S$ and the peak AC output voltage $\hat{v}_x$ can be obtained as follows:

$$\hat{v}_S = \frac{1}{(1-2D_S)(1-D_S)}V_{DC} \text{ and } \hat{v}_x = M\frac{\hat{v}_S}{2} \tag{6.2}$$

Define $B = \frac{1}{(1-2D_S)(1-D_S)}$, the boost factor in the DC side, then the peak AC side can be written as:

$$\hat{v}_x = B\left(M\frac{V_{DC}}{2}\right) \tag{6.3}$$

Now the boosting factor has increased by a factor of $1/(1-D_S)$ compared to that of the original ZSI. Similarly, the steady-state equations can be derived for the diode-assisted extended boost discontinuous current qZSI. Then it is possible to prove that this converter also has the same boosting factor as that of the continuous current topology. Also, the voltage stress on the capacitors are similar, except the voltage across capacitor 3 can be shown to be $V_{c3} = D_S/(1 - D_S)*V_{DC}$. By studying these two topologies, it can be noted that with the discontinuous current topology, the capacitors are subjected to a small voltage stress, and if there is no boosting, then the voltage across them is zero. Also, it is possible to derive the boost factor for topologies shown in Figures 6.2b and 6.3b as $B = \frac{1}{(1-2D_S)(1-D_S)^2}$

6.2.2 Capacitor-Assisted Extended Boost qZSI Topologies

Similar to the previous family of extended boost qZSIs, this section is proposing another family of converters. The difference is now that a much higher boost is achieved with only a simple structural change to the previous topology. Now D_3 is replaced with a capacitor as shown in Figure 6.4. In this context also, two topological variations are derived as continuous current or discontinuous current forms as shown in Figure 6.5.

In the previous scenario, the steady-state relations are derived using the continuous current topology; therefore, in this context the discontinuous current topology is considered. In this case also, the converter's three operating states are simplified into shoot-through and non-shoot-through states.

The simplified circuit diagram is shown in Figure 6.6a. First, consider the non-shoot-through state shown in Figure 6.6b, which is represented with an open switch. As diodes D_1 and D_2 are conducting, the inductors discharge and capacitors get charged. Then by applying KVL, the following steady-state relationships can be observed. $V_{DC} + V_{c3} + V_{c2} + V_{c1} = V_S$ and $V_{DC} + V_{c3} + V_{c4} + V_{c1} = V_S$, $V_{c1} = v_{L1}$, $V_{c2} = v_{L2}$, $V_{c3} = v_{L3}$, $V_{DC} + V_{c3} = V_d$, $V_{c2} = V_{c4}$. Figure 6.6c shows the equivalent circuit diagram for the shoot-through state where it is represented with the closed switch. Both diodes D_1 and D_2 are in blocking state where all the inductors are charged and energy is transferred from the source to inductors or the capacitor to inductors, while capacitors are discharged. Similar

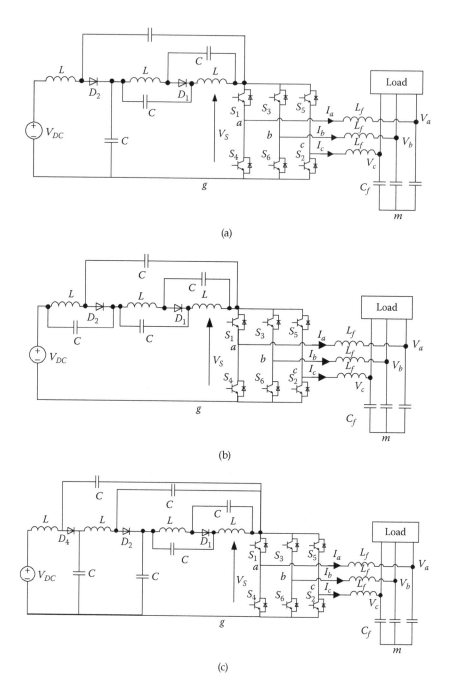

FIGURE 6.5
Capacitor-assisted extended boost qZSIs.

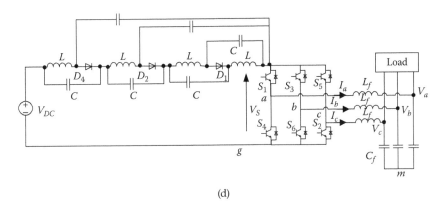

(d)

FIGURE 6.5 (*continued*)
Capacitor-assisted extended boost qZSIs.

relationships can be derived as $V_{DC} + v_{L3} + V_{c4} + V_{c1} = 0$, $V_{DC} + V_{c3} = V_d$, $V_d + V_{L1} + V_{c2} = 0$, $V_d + V_{L2} + V_{c1} = 0$, and $V_S = 0$. Considering the fact that the average voltage across the inductors is zero, the following relations can be derived:

$$V_d = \frac{1-2D_s}{1-3D_s}V_{DC} \quad \text{and} \quad V_{C1} = V_{C2} = V_{C3} = V_{C4} = \frac{D_s}{1-2D_s}V_d = \frac{D_s}{1-3D_s}V_{DC} \quad (6.4)$$

Then, from the above equations, the peak voltage across the inverter $\hat{v}_S$ can be obtained as follows:

$$\hat{v}_S = \frac{1}{1-3D_S}V_{DC} \quad (6.5)$$

Similar equations can be derived for the continuous current topology. Now the difference would be the continuity of source current and the difference in voltage across the. The voltage across the C_3 can be obtained as $V_{C3} = V_d$. Now the voltage across the capacitor is much larger than with discontinuous current topology. Similarly, it is possible to derive the boost factor for topologies shown in Figure 6.5c and 6.5d as $B = \frac{1}{1-4D_S}$.

6.2.3 Simulation Results

Extensive simulation studies are performed on the open-loop configuration of all proposed topologies in Matlab/Simulink® using the modulation method proposed [1]. However, due to space limitations, only a few results are presented. This would validate the operation of diode-assisted and capacitor-assisted topologies as well as continuous current and discontinuous current topologies. Here, three cases are simulated. In all three cases, the input voltage is kept constant at 240 V and a three-phase load of 9.7 Ω resistor bank is used. All DC side capacitors are 1000 μF, and inductors are 3.5 mH. The AC side

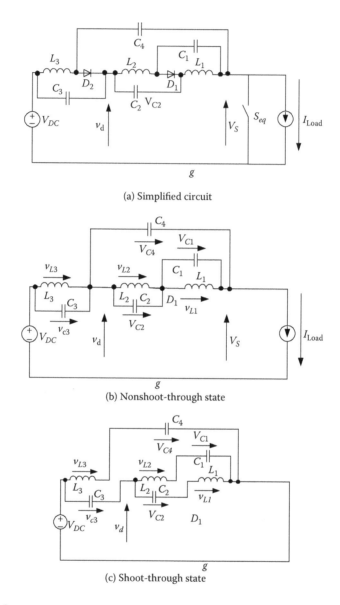

(a) Simplified circuit

(b) Nonshoot-through state

(c) Shoot-through state

FIGURE 6.6
Simplified diagram of capacitor-assisted extended boost continuous current qZSI.

second-order filter is used with a 10 μF capacitor and a 7 mH inductor. In all three cases, the converter is operated with zero boosting in the beginning and, at $t = 250$ ms, the shoot-through is increased to 0.25 while the modulation index kept constant at 0.7. Figures 6.7–6.9 show the simulation results corresponding to topologies shown in Figures 6.2a, 6.3a, and 6.5b. From these figures, it is

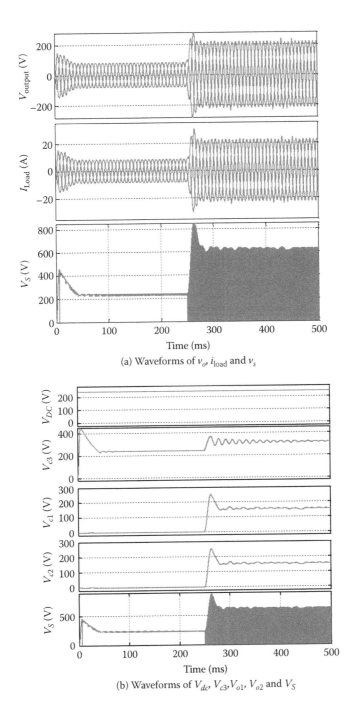

(a) Waveforms of v_o, i_{load} and v_s

(b) Waveforms of V_{dc}, V_{c3}, V_{o1}, V_{o2} and V_S

FIGURE 6.7
Simulation results for diode-assisted extended boost continuous current qZSI.

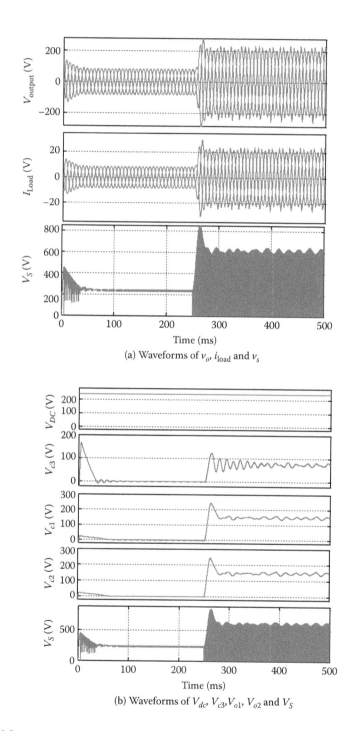

(a) Waveforms of v_o, i_{load} and v_s

(b) Waveforms of V_{dc}, V_{c3}, V_{o1}, V_{o2} and V_S

FIGURE 6.8
Simulation results for diode-assisted extended boost discontinuous current qZSI.

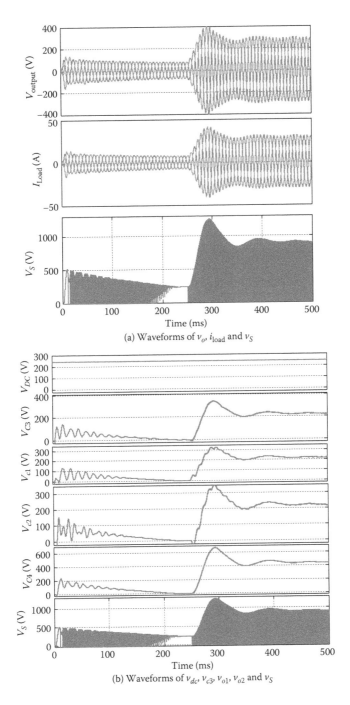

(a) Waveforms of v_o, i_{load} and v_S

(b) Waveforms of v_{dc}, v_{c3}, v_{o1}, v_{o2} and v_S

FIGURE 6.9

Simulation results for capacitor-assisted extended boost discontinuous current qZSI.

possible to note that in the first two cases equal boosting is achieved and the difference is the voltage across $VC3$. This agrees with the theoretical finding. From Figure 6.9, it can be noted that with the capacitor-assisted topology a much higher boosting can be achieved with the same shoot-through value; also voltages across all four capacitors are equal and agree with the equations derived in the previous section. A comprehensive set of simulation results will be presented in the full paper based on this chapter.

References

1. Gajanayake, C. J. and Luo, F. L. 2009. Extended boost Z-source inverters. *Proc. IEEE ECCE'2009*, pp. 368–373.
2. Gajanayake,C. J., Vilathgamuwa, D. M. and Loh, P. C., 2007. Development of a comprehensive model and a multiloop controller for Z-source inverter DG systems. *IEEE Trans. Ind. Electron.*, pp.2352–2359.
3. Anderson, J. and Peng, F. Z. 2008. Four quasi-Z-Source inverters. *Proc. IEEE PESC'2008*, pp. 2743–2749.
4. Luo, F. L. and Ye, H. 2005. *Advanced DC/DC Converters*. Boca Raton, FL: CRC Press.
5. Luo, F. L. and Ye, H. 2005. *Essential DC/DC Converters*, Boca Raton, FL: Taylor & Francis.
6. Luo, F. L. 1999. Positive output Luo-converters: Voltage lift technique. *IEE Proc. Electric Power Applicat.*, pp. 415–432.
7. Luo, F. L. 1999. Positive output Luo-converters: Voltage lift technique. *IEE Proc. Electric Power Applicat.*, pp. 208–224.
8. Ortiz-Lopez, M. G., Leyva-Ramos, J. E., Carbajal-Gutierrez, E., and Morales-Saldana, J. A. 2008. Modelling and analysis of switch-mode cascade converters with a single active switch. *Power Electronics, IET*, pp. 478–487.

7

Soft-Switching DC/AC Inverters

The soft-switching technique has been implemented in DC/DC conversion for more than 25 years. We also would like to introduce this technique in DC/AC inverters in this chapter.

There are numerous ways of implementing soft-switching methods, such as zero voltage switching (ZVS) and zero current switching (ZCS), to reduce the switching losses and to increase efficiency for different multilevel inverters. For the cascaded multilevel inverter, because each inverter cell is a two-level circuit, the implementation of soft switching is not at all different from that of conventional two-level inverters. For capacitor- or diode-clamped inverters, however, the choices of soft switching circuit involve different circuit combinations. Although zero current switching is possible, most approaches in the literature propose zero voltage switching types, including auxiliary resonant commutated pole (ARCP), coupled inductor with zero voltage transition (ZVT), and their combinations.

7.1 Notched DC Link Inverters for Brushless DC Motor Drive

The brushless DC motor (BDCM) has been widely used in industrial applications because of its low inertia, fast response, high power density, and high reliability, because of which it is maintenance-free. It exhibits the operating characteristics of a conventional commutated DC permanent magnet motor but eliminates the mechanical commutator and brushes. Hence, many problems associated with brushes are eliminated, such as radio-frequency interference and sparking, which is the potential source of ignition in an inflammable atmosphere. It is usually supplied by a hard-switching PWM inverter, which normally has low efficiency since the power losses across the switching devices are high. In order to reduce the losses, many soft-switching inverters have been designed [1].

The soft-switching operation of power inverters has attracted much attention in recent decades. In electric motor drive applications, soft-switching inverters are usually classified into three categories, namely, resonant pole inverters, resonant DC link inverters, and resonant AC link inverters [2]. The resonant pole inverter has the disadvantage of containing a large number of additional components, in comparison to other hard- and soft-switching inverter topologies. The resonant AC link inverter is not suitable for BDCM drivers.

In medium-power applications, the resonant DC link concept [3] offered a first practical and reliable way to reduce commutation losses and to eliminate individual snubbers. Thus, it allows high operating frequencies and improved efficiency. In this inverter, it is quite simple to get the zero voltage switching (ZVS) condition of the six main switches just by adding one auxiliary switch. However, the inverter has the drawbacks of high voltage stress of the switches and high voltage ripple of the DC link, the frequency of the inverter being related to the resonant frequency. Furthermore, the inductor power losses of the inverter are also considerable as current always flows in the inductor. In order to overcome the drawback of high voltage stress of the switches, the actively clamped resonant DC link inverter was introduced [4–7]. The control scheme of the inverter is exceedingly complex and the output contains subharmonics, which, in some cases, cannot be accepted. These inverters still do not overcome the drawback of high inductor power losses.

In order to generate voltage notches of the DC link at controllable instants and reduce the power losses of the inductor, several quasi-parallel resonant schemes were proposed [8–10]. As a dwell time is generally required after every notch, severe interference occurs, mainly in multiphase inverters, appreciably worsening the modulation quality. A novel DC-rail parallel resonant zero voltage transition (ZVT) voltage source inverter [11] was introduced that overcame many of the drawbacks mentioned earlier. However, it requires two ZVTs per PWM cycle, which lowers the output and limits the switch frequency of the inverter.

On the other hand, the majority of soft-switching inverters proposed in recent years have been aimed at induction motor drive applications. So it is necessary to research the novel topology of the soft-switching inverter and the special control circuit for BDCM drive systems. This chapter proposes a novel resonant DC link inverter for the BDCM drive system that can generate voltage notches of the DC link at controllable instants and widths. And the inverter possesses the advantages of low switching power loss, low inductor power loss, low voltage ripple of the DC link, low device voltage stress, and a simple control scheme.

The construction of the soft-switching inverter is shown in Figure 7.1. At the front is an uncontrolled rectifier for DC supply. The input AC supply can be single phase for low/medium power or three phases for medium/high power. It contains a resonant circuit, a conventional and control circuit. The resonant circuit contains three auxiliary switches (one IGBT and two fast switching thyristors), a resonant inductor, and a resonant capacitor. All auxiliary switches work under ZVS or zero current switching (ZCS) condition. It generates voltage notches of the DC link to guarantee that the main switches S_1–S_6 of the inverter will operate in ZVS condition. The fast-switching

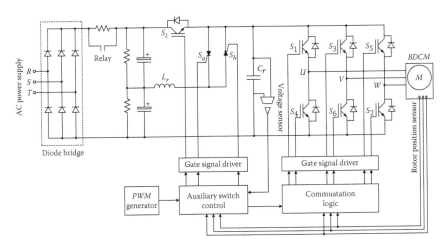

FIGURE 7.1
The construction of soft-switching for the BDCM drive system.

thyristor is the appropriate device as an auxiliary switch. We need not control the turn-off of a thyristor, because it has higher surge current capability than any other power semiconductor switch.

7.1.1 Resonant Circuit

The resonant circuit consists of three auxiliary switches, one resonant inductor, and one resonant capacitor. The auxiliary switches are controlled at a certain instant for resonance between inductor and capacitor. Thus, the voltage of the DC link reaches zero temporarily (voltage notch), and the main switches of the inverter reach assumed ZVS condition for commutation.

Since the resonant process is very short, the load current can be assumed constant. The equivalent resonant circuit is shown in Figure 7.2.

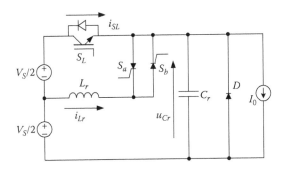

FIGURE 7.2
The equivalent resonant circuit.

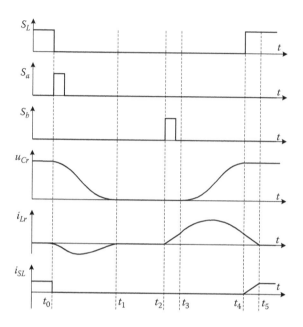

FIGURE 7.3
Some waveforms of the equivalent circuit.

The corresponding waveforms of the auxiliary switches gate signal, resonant capacitor voltage (u_{Cr}), inductor current (i_{Lr}), and current of switch S_L (i_{SL}) are shown in Figure 7.3. The operation of the ZVT process can be divided into six modes.

Mode 0 (as shown in Figure 7.4a) $0 < t < t_0$. Its operation is the same as that of the conventional inverter. Current flows from the DC source through S_L to the load. The voltage across C_r (u_{Cr}) is equal to the supply voltage (V_s). The auxiliary switches S_a and S_b are in the off state.

Mode 1 (as shown in Figure 7.4b) $t_0 < t < t_1$. At the instant of phase current commutation or when the PWM signal is flipped from 1 to 0, thyristor S_a is fired (ZCS turns on due to L_r) and IGBT S_L is turned off (ZVS turns off due to C_r) at the same time. Capacitor C_r resonates with inductor L_r, and the voltage across capacitor C_r is decreased. Redefining the initial time, we have the equation

$$\begin{cases} u_{Cr}(t) + R_{Lr}i_{Lr}(t) + L_r\dfrac{di_{Lr}(t)}{dt} = \dfrac{V_S}{2} \\ I_0 - i_{Lr}(t) + C_r\dfrac{du_{cr}(t)}{dt} = 0 \end{cases} \qquad (7.1)$$

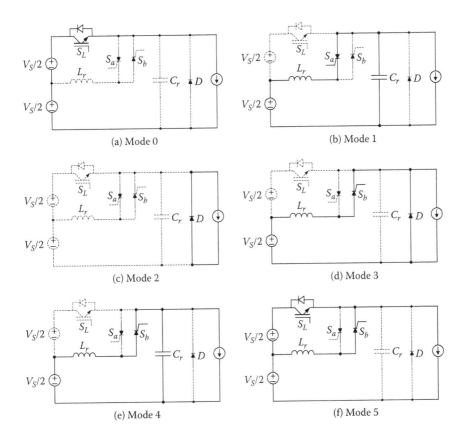

FIGURE 7.4
Operation mode of the zero voltage switching process.

R_{Lr} is the resistance of inductor L_r, I_0 is the load current, V_S is the DC power supply voltage, with initial condition $u_{cr}(0) = V_S$, $i_{Lr}(0) = 0$. Solving Equation (7.1), we get

$$
\begin{cases}
u_{Cr}(t) = \left(\dfrac{V_S}{2} - R_{Lr}I_0\right) + \left(\dfrac{V_S}{2} - R_{Lr}I_0\right)e^{-\frac{t}{\tau}}\cos(\omega t) \\[2ex]
\qquad + \dfrac{1}{L_r C_r \omega}e^{-\frac{t}{\tau}}\left(\dfrac{1}{4}R_{Lr}C_r V_S - L_r I_0 + \dfrac{1}{2}R_{Lr}^2 C_r I_0\right)\sin(\omega t) \quad (7.2) \\[2ex]
i_{Lr}(t) = I_0 - I_0 e^{-\frac{t}{\tau}}\cos(\omega t) - \dfrac{V_S + R_{Lr}I_0}{2L_r\omega}e^{-\frac{t}{\tau}}\sin(\omega t)
\end{cases}
$$

where

$$\tau = \frac{2L_r}{R_{Lr}}, \quad \omega = \sqrt{\frac{1}{L_r C_r} - \frac{1}{\tau^2}}.$$

As the resonant frequency is very high (several hundred kHz), $\omega L_r \gg R_{Lr}$, and the resonant inductor resistance R_{Lr} can be neglected. Then Equation (7.2) can be simplified as

$$\begin{cases} u_{Cr}(t) = \dfrac{V_S}{2} - I_0 \sqrt{\dfrac{L_r}{C_r}} \sin\left(\dfrac{1}{\sqrt{L_r C_r}} t\right) + \dfrac{V_S}{2} \cos\left(\dfrac{1}{\sqrt{L_r C_r}} t\right) \\[4mm] i_{Lr}(t) = I_0 - I_0 \cos\left(\dfrac{1}{\sqrt{L_r C_r}} t\right) - \dfrac{V_S}{2} \sqrt{\dfrac{C_r}{L_r}} \sin\left(\dfrac{1}{\sqrt{L_r C_r}} t\right) \end{cases} \tag{7.3}$$

That is,

$$\begin{cases} u_{Cr}(t) = \dfrac{V_S}{2} + K \cos(\omega_r t + \alpha) \\[4mm] i_{Lr}(t) = I_0 - K \sqrt{\dfrac{C_r}{L_r}} \sin(\omega_r t + \alpha) \end{cases} \tag{7.4}$$

where

$$K = \sqrt{\frac{V_S^2}{4} + \frac{I_0^2 L_r}{C_r}}, \quad \omega_r = \sqrt{\frac{1}{L_r C_r}}, \quad \alpha = tg^{-1}\left(\frac{2I_0}{V_S}\sqrt{\frac{L_r}{C_r}}\right)$$

Letting $u_{Cr}(t) = 0$, we get

$$\Delta T_1 = t_1 - t_0 = \frac{\pi - 2\alpha}{\omega_r} \tag{7.5}$$

$i_{Lr}(t_1)$ is zero at the same time. Then thyristor S_a turns itself off.

Mode 2 (as shown in Figure 7.4c) $t_1 < t < t_2$. None of the auxiliary switches is fired, and the voltage of the DC link (u_{Cr}) is zero. The main switches of the inverter can now be either turned on or turned off under ZVS condition during the interval. Load current flows through the freewheeling diode D.

Mode 3 (as shown in Figure 7.4d) $t_2 < t < t_3$. As the main switches have turned on or off, thyristor S_b is fired (ZCS turns on due to L_r), and i_{Lr} starts to build up linearly in the auxiliary branch. The current in the freewheeling diode D begins to fall linearly. The load current is slowly diverted from the freewheeling diodes to the resonant branch. But u_{Cr} is still equal to zero. We have

$$\Delta T_2 = t_3 - t_2 = \frac{2I_0 L_r}{V_S} \tag{7.6}$$

At t_3, i_{Lr} equals the load current I_0, and the current through the diode becomes zero. Thus, the freewheeling diode turns off under zero-current condition.

Mode 4 (as shown in Figure 7.4e) $t_3 < t < t_4$. i_{Lr} is increased continuously from I_0 and u_{Cr} is increased from zero when the freewheeling diode D is turned off. Redefining the initial time, we can get the same equation as Equation (7.1). But the initial condition is $u_{Cr}(0) = 0$, $i_{Lr}(0) = I_0$. Neglecting the inductor resistance and solving the equation, we get

$$\begin{cases} u_{Cr}(t) = \dfrac{V_S}{2} - \dfrac{V_S}{2}\cos\left(\dfrac{1}{\sqrt{L_r C_r}}t\right) \\[4mm] i_{Lr}(t) = I_0 + \dfrac{V_S}{2}\sqrt{\dfrac{C_r}{L_r}}\sin\left(\dfrac{1}{\sqrt{L_r C_r}}t\right) \end{cases} \tag{7.7}$$

That is,

$$\begin{cases} u_{Cr}(t) = \dfrac{V_S}{2}[1 - \cos(\omega_r t)] \\[4mm] i_{Lr}(t) = I_0 + \dfrac{V_S}{2}\sqrt{\dfrac{C_r}{L_r}}\sin(\omega_r t) \end{cases} \tag{7.8}$$

When

$$\Delta T = t_4 - t_3 = \frac{\pi}{\omega_r} \tag{7.9}$$

$u_{Cr} = E$, IGBT S_L is fired (ZVS turn on), and $i_{Lr} = I_0$ again. The peak inductor current can be derived from Equation (7.8) as

$$i_{Lr-m} = I_0 + \frac{V_S}{2}\sqrt{\frac{C_r}{L_r}} \tag{7.10}$$

Mode 5 (as shown in Figure 7.4f) $t_4 < t < t_5$. When the DC link voltage is equal to the supply voltage, auxiliary switch S_L is turned on (ZVS turns on due to C_r). i_{Lr} decreases linear from I_0 to zero at t_5, and the thyristor S_b turns itself off.

Then the device returns to mode 0 again. The operation principle of the other procedure is the same as for a conventional inverter.

7.1.2 Design Considerations

The design of the resonant circuit is to determine the resonant capacitor C_r, resonant inductor L_r, and the switching instants of auxiliary switches S_a, S_b, and S_L. It is assumed that the inductance of BDCM is much higher than the resonant inductance L_r. From the analysis presented previously, the design considerations can be summarized as follows:

The auxiliary switch S_L works under ZVS condition, and the voltage stress is the DC power supply voltage V_S. The current flow through it is load current. The auxiliary switches S_a and S_b work under ZCS condition, the voltage stress is $V_S/2$, and the peak current flow through them is i_{Lr-m}. As the resonant auxiliary switches S_a and S_b carry peak current only during switch transitions, they can be rated for a lower continuous current rating.

The resonant period is expressed as $T_r = 1/f_r = 2\pi\sqrt{L_r C_r}$; for a high-switching frequency inverter, T_r should be as short as possible. For getting the expected T_r, the resonant inductor and capacitor values have to be selected. The first component to be designed is the resonant inductor. A small inductance value can ensure small T_r, but the rising slope of the inductor current $di_{Lr}/dt = V_S/2L_r$ should be small enough to guarantee freewheeling diode turn-off. For a 600V to 1200V power diode, the reverse recovery time is about 50 to 200 ns, the rule to select an inductor is given by

$$\frac{di_{Lr}}{dt} = \frac{V_S}{2L_r} = 75 \sim 150 A/\mu s \qquad (7.11)$$

Certainly, inductance is as high as possible. This implies that a high inductance value is necessary. Thus, an optimum value of the inductance has to be chosen that would reduce the inductor current rise slope, while keeping T_r small enough.

The capacitance value is inversely proportional to the ascending or descending slope of the DC link voltage. This means the capacitance is as high as possible for switch S_L to get ZVS condition, but as the capacitance increases, more and more energy is stored in it. This energy should be charged or discharged via a resonant inductor; with high capacitance, the peak value of inductor current will be high. The peak value of i_{Lr} should be limited to twice the peak load current. From Equations (7.3)–(7.10), we obtain

$$\sqrt{\frac{C_r}{L_r}} \leq \frac{2I_{0max}}{V_S} \qquad (7.12)$$

Thus, an optimum value of the capacitance has to be chosen that would limit the peak inductor current, while the ascending or descending slope of the DC link voltage is low enough.

7.1.3 Control Scheme

When the duty of PWM is 100%, that is, there is no PWM, the main switches of the inverter work under the commutation frequency. When it is time to commutate the phase current of the BDCM, we control the auxiliary switches S_a, S_b, and S_L, and resonance occurs between L_r and C_r. The voltage of the DC link reaches zero temporarily; thus, ZVS condition of the main switches is obtained. When the duty of the PWM is less than 100%, the auxiliary switch S_L works as a chopper. The main switches of the inverter do not switch on within a PWM cycle when the phase current need not commutate. It has the benefit of reducing the phase current drop when the PWM is off. The phase current is commutated when the DC link voltage becomes zero. So there is only one DC link voltage notch per PWM cycle. It is very important, especially for very low or very high duty of PWM, where the interval between two voltage notches is very short or even overlapped which will limit the tuning range.

The commutation logical circuit of the system is shown in Figure 7.5. It is similar to the conventional BDCM commutation logical circuit except that six D flip-flops are added to the output. Thus, the gate signal of the main

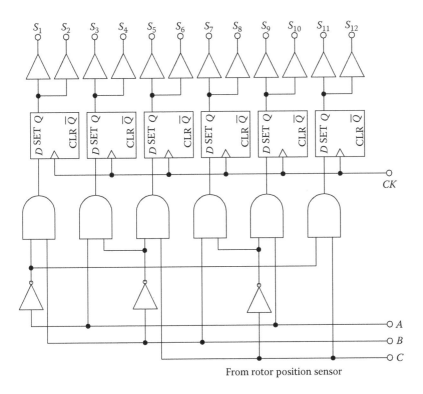

FIGURE 7.5
Commutation logical circuit for main switches.

switches is controlled by the synchronous pulse CK, which will be mentioned later, and the commutation can be synchronized with the auxiliary switches' control circuit. The operation of the inverter can be divided into PWM operation and non-PWM operation.

7.1.3.1 Non-PWM Operation

When the duty of PWM is 100%, that is, there is no PWM, the whole ZVT process (mode 1–mode 5) occurs when the phase current commutation is under way. The control scheme for the auxiliary switches in this operation is illustrated in Figure 7.6a. When mode 1 begins, the pulse signal for thyristor S_a is

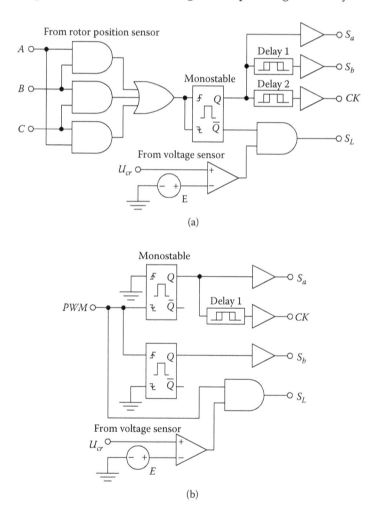

FIGURE 7.6
Control scheme for the auxiliary switches in (a) non-PWM operation and (b) PWM operation.

generated by a monostable flip-flop, and the gate signal for IGBT S_L drops to a low level (i.e., turns off the S_L) at the same time. Then the pulse signal for thyristor S_b and the synchronous pulse CK can be obtained after two short delays (delay 1 and delay 2, respectively). Obviously, delay 1 is longer than delay 2. Pulse CK is generated during mode 2 when the voltage of DC link is zero and the main switches of the inverter get ZVS condition. Then modes 3, 4, and 5 occur and the voltage of DC link is increased to that of the supply again.

7.1.3.2 PWM Operation

In this operation, the auxiliary switch S_L works as a chopper, but the main switches of the inverter do not turn on or off within a single PWM cycle when the phase current need not commutate. The load current is commutated when the DC link voltage becomes zero, that is, when the PWM signal is 0 (as the PWM cycle is very short, it does not affect the operation of the motor). The control scheme for the auxiliary switches in PWM operation is illustrated in Figure 7.6b.

- When the PWM signal is flipped from 1 to 0, mode 1 begins, the pulse signal for thyristor S_a is generated, and the gate signal for IGBT S_L drops to a low level. But the voltage of the DC link does not increase until the PWM signal is flipped from 0 to 1. Pulse CK is generated during mode 2.
- When the PWM signal is flipped from 0 to 1, mode 3 begins, and the pulse signal for thyristor S_b is generated (mode 3). Then when the voltage of the DC link is increased to E (voltage of supply), the gate signal for IGBT S_L is flipped to a high level (modes 4 and 5).

Thus, only one ZVT occurs per PWM cycle: modes 1 and 2 for PWM turned off, modes 3, 4, and 5 for PWM turned on. And the switching frequency would not be greater than the PWM frequency.

Normally, a drive system requires a speed or position feedback signal to get high speed or position precision and be less susceptible to disturbance of load and power supply. A speed feedback signal can be derived from a tachometer-generator, optical encoder, or resolver, or it can be derived from the rotor position sensor. The quadrature encoder pulse (QEP) is a standard digital speed or position signal and can be input to many devices (e.g., special DSP for drive system TMS320C24x has a QEP receive circuit). The QEP can be derived from the rotor position sensor of a BDCM easily. The converter digital circuit and interesting waveforms are shown in Figure 7.7. Some single-chip computers have a digital counter and may require only direction and pulse signals; thus, the converter circuit can be simplified. The circuit can be implemented with a complex programmer logical device and only occupies part of one chip. The circuit can also be implemented by a gate array logic IC (e.g., 16V8) and a D flip-flop IC (e.g., 74LS74). With the circuit, a high-precision

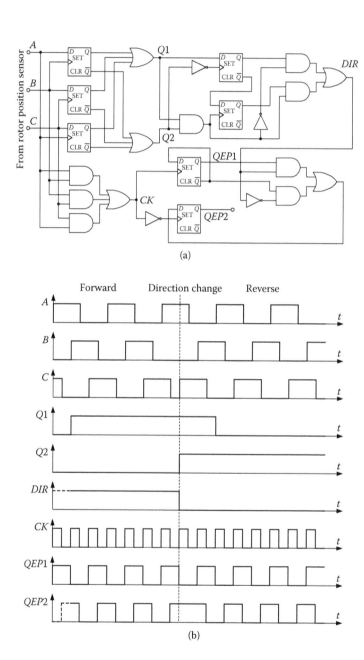

FIGURE 7.7
Circuit of derived QEP from Hall signal and waveforms.

speed or position signal can be obtained when the motor speed is high or the drive system has a high ratio speed reduction mechanism. In high-performance systems, the rotor position sensor may be a resolver or optical encoder, with special-purpose decoding circuitry. At this level of control sophistication, it is possible to fine-tune the firing angles and the PWM control as a function of speed and load, to improve various aspects of performance such as efficiency, dynamic performance, or speed range.

7.1.4 Simulation and Experimental Results

The proposed topology is verified by Psim simulation software. The schematic circuit of the soft-switching inverter is shown in Figure 7.8. The left bottom of the figure is the auxiliary switches' gate signal generator circuit (see Figure 7.6), which is made up of a monostable flip-flop, delay, and logical gate. The gate signals of auxiliary switches S_a and S_b in PWM and no-PWM operation modes are combined by an OR gate. The gate signal of S_L in the two operation modes is combined by an AND gate, and the synchronous signal (CK) is combined by a date selector. The middle bottom of the diagram shows the commutation logical circuit of the BDCM (see Figure 7.5); it is synchronized (by CK) with the auxiliary switches' control circuit.

Waveforms of DC link voltage u_{Cr}, resonant inductor current i_{Lr}, BDCM phase current, inverter output line-line voltage, and gate signal of the auxiliary switches are shown in Figure 7.9. The resonant inductor L_r has an inductance of 10 μH, and the resonant capacitor C_r has a capacitance of 0.047 μF, so the period of the resonant circuit is about 4 μs. The frequency of the PWM is 20 kHz. From the figure, we can see that the output of the simulation matches the theoretical analysis. The waveforms in Figure 7.9b, d, e, f, g, and h are the same as in Figure 7.10.

In order to verify the theoretical analysis and simulation results, the proposed soft-switching inverter was tested on an experimental prototype rated as follows:

DC link voltage	240 V
Power of BDCM	3.3 Hp
Switching frequency	10 kHz

A polyester capacitor of 47 nF, 1500 V, was adopted as the DC link resonant capacitor C_r. The resonant inductor had an inductance of 6 μH/20 A, with a ferrite core. The design of the auxiliary switches' control circuit was referenced from Figure 7.8. The monostable flip-flop can be implemented with IC 74LS123, the delay can be implemented by a Schmitt trigger and an RC circuit, and the logical gate can be replaced by a programmable logical device to reduce the number of ICs.

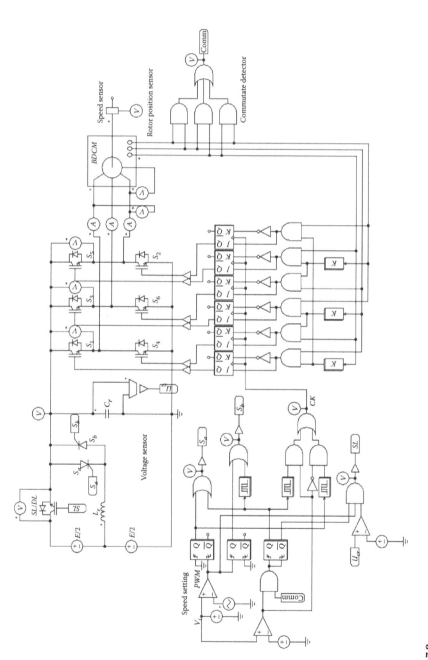

FIGURE 7.8
Schematic circuit of the drive system for Psim simulation.

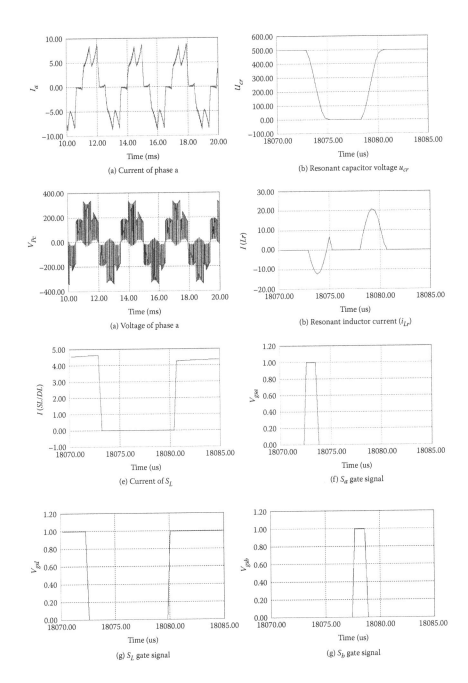

(a) Current of phase a

(b) Resonant capacitor voltage u_{cr}

(a) Voltage of phase a

(b) Resonant inductor current (i_{Lr})

(e) Current of S_L

(f) S_a gate signal

(g) S_L gate signal

(g) S_b gate signal

FIGURE 7.9
Simulation results.

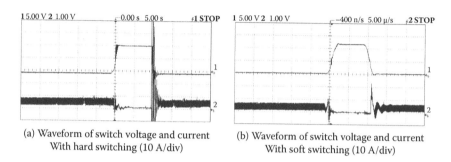

(a) Waveform of switch voltage and current With hard switching (10 A/div)

(b) Waveform of switch voltage and current With soft switching (10 A/div)

FIGURE 7.10

Voltage and current waveforms of switch S_L in hard-switching and soft-switching inverters.

The waveforms of voltage across the switch and current under hard switching and soft switching are shown in Figures 7.10a and 7.10b, respectively. All the voltage signals come from differential probes, and there is a gain of 20. For voltage waveforms, 5.00 V/div = 100 V/div, which is the same for Figure 7.11. It can also be seen that there is a considerable overlap between the voltage and current waveforms under hard switching. The overlap is much less with soft switching.

The key waveforms with soft switching inverter are shown in Figure 7.11. The default scale is as follows: DC link voltage: 100 V/div, current: 20 A/div. The default switching frequency is 10 kHz. The DC link voltage is

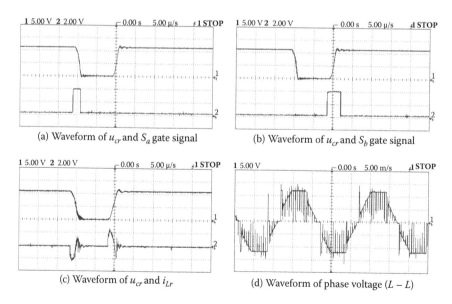

(a) Waveform of u_{cr} and S_a gate signal

(b) Waveform of u_{cr} and S_b gate signal

(c) Waveform of u_{cr} and i_{Lr}

(d) Waveform of phase voltage $(L - L)$

FIGURE 7.11

Experiment waveforms.

fixed at 240 V. These experimental waveforms are similar to the simulation waveforms in Figure 7.9.

7.2 Resonant Pole Inverter

The resonant pole inverter is a soft-switching DC/AC inverter circuit and is shown in Figure 7.12. Each resonant pole comprises a resonant inductor and a pair of resonant capacitors at each phase leg. These capacitors are directly connected in parallel to the main inverter switches in order to achieve zero voltage switching (ZVS) condition. In contrast to the resonant DC link inverter, the DC link voltage remains unaffected during the resonant transitions. These transitions occur separately at each resonant pole when the corresponding main inverter switch needs switching. Therefore, the main switches in the inverter phase legs can switch independently of each other and choose the commutation instant freely. Moreover, there is no additional main conduction path switch. Thus, the normal operation of the resonant pole inverter is precisely the same as that of the conventional hard-switching inverter [12].

The auxiliary resonant commutated pole (ARCP) inverter [13] and the ordinary resonant snubber inverter [14] provide a ZVS condition without increasing the device voltage and current stress. These inverters are able to achieve real PWM control. However, they require a stiff DC link capacitor bank that is center-taped to accomplish commutation. The center voltage of the DC link is susceptible to drift, which may affect the operation of the resonant circuit. The resonant transition inverter [15,16] uses only one auxiliary switch, whose switching frequency is much higher than that applied to the main switches. Thus, it will limit the switching frequency of the inverter. Furthermore, the three resonant branches of the inverter work together and will be affected by each other. A Y-configured resonant snubber inverter [17] has a floating

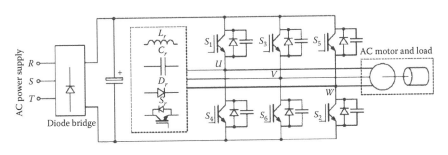

FIGURE 7.12
Resonant pole inverter.

neutral voltage that may cause overvoltage failure of the auxiliary switches. A delta (Δ)-configured resonant snubber inverter [18] avoids the floating neutral voltage and is suitable for multiphase operation without circulating currents between the off-state branch and its corresponding output load. However, the inverter requires three inductors and six auxiliary switches.

Moreover, resonant pole inverters have been applied in induction motor drive applications. They are usually required to change two phase switch states at the same time to obtain a resonant path. It is not suitable for a BDCM drive system as only one switch is needed to change the switching state in a PWM cycle. The switching frequency of three upper switches (S_1, S_3, S_5) is different from that of the three lower switches (S_2, S_4, S_6) in an inverter for a BDCM drive system. All the switches have the same switching frequency in a conventional inverter for induction motor applications. Therefore, it is necessary to develop a novel topology for the soft-switching inverter and a special control circuit for BDCM drive systems. This chapter presents a specially designed resonant pole inverter that is suitable for BDCM drive systems and is easy to apply in industry. In addition, this inverter possesses the following advantages: low switching power losses, low inductor power losses, low switching noise, and simple control scheme.

7.2.1 Topology of Resonant Pole Inverter

A typical controller for a BDCM drive system [19] is shown in Figure 7.13.

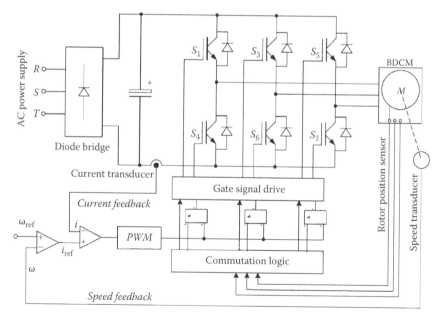

FIGURE 7.13
Typical controller for the BDCM drive system.

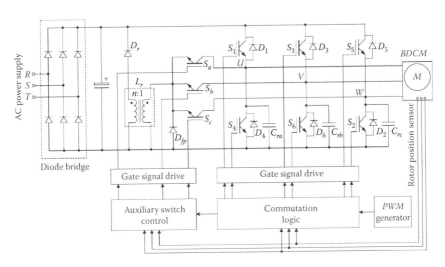

FIGURE 7.14
Structure of the resonant pole inverter for the BDCM drive system.

The rotor position can be sensed by a Hall effect sensor or a slotted optical disk, providing three square-waves with phase shift in 120°. These signals are decoded by a combinatorial logic to provide the firing signals for 120° conduction on each of the three phases. The basic forward control loop is voltage control implemented by PWM (voltage reference signal compared with triangular wave or generated by microprocessor). The PWM is applied only to the lower switches. This not only reduces the current ripple but also avoids the need for wide bandwidth in the level-shifting circuit that feeds the upper switches. The three upper switches work under commutation frequency (typically, several hundred Hz) and the three lower switches work under PWM frequency (typically tens of kHz). So it is not important that the three upper switches work under soft-switching condition. The switching power losses can be reduced significantly and the auxiliary circuit would be simpler if only the three lower switches work under soft-switching condition. Thus, a specially designed resonant pole inverter for the BDCM drive system was introduced for this purpose. The structure of the proposed inverter is shown in Figure 7.14.

The system contains a diode bridge rectifier, a resonant circuit, a conventional three-phase inverter, and control circuitry. The resonant circuit consists of three auxiliary switches (S_a, S_b, S_c), one transformer with turn ratio 1: n, and two diodes D_{fp} and D_r. Diode D_{fp} is connected in parallel to the primary winding of the transformer, and diode D_r is connected in series with the secondary winding across the DC link. There is one snubber capacitor connected in parallel to each lower switch of the phase leg. The snubber capacitor resonates with the primary winding of the transformer. The emitters of the three auxiliary switches are connected together. Thus, the gate drive of these auxiliary switches can use one common output DC power supply.

In a whole PWM cycle, the three lower switches (S_2, S_4, S_6) can be turned off in the ZVS condition as the snubber capacitors (C_{ra}, C_{rb}, C_{rc}) can slow down the voltage rise rate. The turn-off power losses can be reduced, and the turn-off voltage spike is eliminated. Before turning on the lower switch, the corresponding auxiliary switch (S_a, S_b, S_c) must be turned on. The snubber capacitor is then discharged, and the lower switches reach the ZVS condition. During phase current commutation, the switching state is changed from one lower switch to another. For example, turn off S_6 and turn on S_2, S_6 can be turned off directly in the ZVS condition, turning on auxiliary switch S_c to discharge the snubber capacitor C_r then switch S_2 can achieve the ZVS condition. During phase current commutation, if the switching state is changed from one upper switch to another upper switch, the operation is the same as that of the hard-switching inverter, as the switching power losses of the upper switches are much smaller than those of the lower switches.

7.2.2 Operation Principle

For the sake of convenience, to describe the operation principle, we investigate the period of time when switch S_1 is always turned on, when switch S_6 works under PWM frequency, and when other main inverter switches are tuned off. Since the resonant transition is very short, it can be assumed that the load current is constant. The equivalent circuit is shown in Figure 7.15. V_S is the DC link voltage, i_{Lr} is the transformer primary winding current, u_{S6} is the voltage drop across the switch S_6 (i.e., snubber capacitor C_{rb} voltage), and I_O is the load current. The waveforms of the switches' (S_6, S_b) gate signal, PWM signal, main switch S_6 voltage drop (u_{S6}), and the transformer primary winding current (i_{Lr}) are shown in Figure 7.16, and the details will now be explained. Accordingly, at the instant $t_0 - t_6$, the operation of one switching cycle can be divided into seven modes.

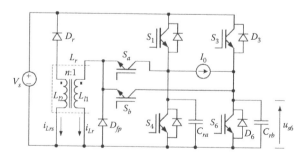

FIGURE 7.15
Equivalent resonant circuit.

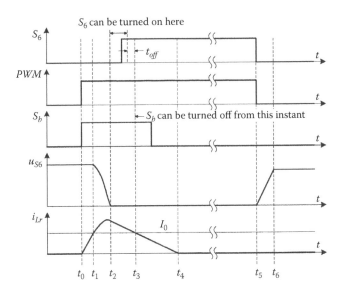

FIGURE 7.16
Key waveforms of the equivalent circuit.

Mode 0 [shown in Figure 7.17a] $0 < t < t_0$: After the lower switch S_6 is turned off, load current flows through the upper freewheeling diode D_3, the voltage drop u_{S6} (i.e., snubber capacitor C_{rb} voltage) across the switch S_6 is the same as that of the DC link voltage. The auxiliary resonant circuit does not operate.

Mode 1 [shown in Figure 7.17b] $t_0 < t < t_1$: If the switch S_6 is turned on directly, the capacitor discharge surge current will also flow through switch S_6; thus, switch S_6 may face the risk of a second breakdown. The energy stored in the snubber capacitor must be discharged ahead of time. Thus, auxiliary switch S_b is turned on (ZCS turns on as the current i_{Lr} cannot change suddenly due to the transformer inductance). As the transformer primary winding current i_{Lr} begins to increase, the current flowing through the freewheeling diode decays. The secondary winding current i_{Lrs} also begins to conduct through diode D_r to the DC link. Both terminal voltages of the primary and secondary windings are equal to the DC link voltage V_S. By neglecting the resistances of the windings and using the transformer equivalent circuit (referred to the primary side) [20], we get

$$V_S = L_{l1} \frac{di_{Lr}(t)}{dt} + a^2 L_{l2} \frac{d[i_{Lrs}(t)/a]}{dt} + aV_S \qquad (7.13)$$

where L_{l1} and L_{l2} are the primary and secondary winding leakage inductance, respectively, and is the transformer turn ratio 1:n. The transformer has

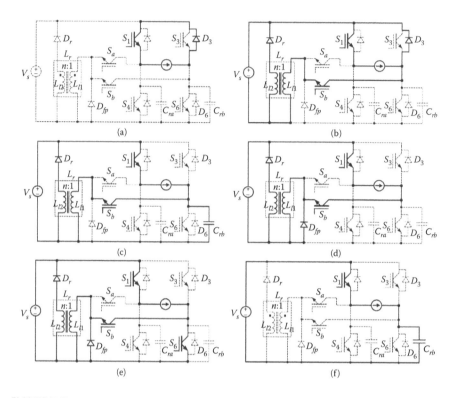

FIGURE 7.17
Operation modes of the resonant pole inverter: (a) Mode 0, (b) Mode 1, (c) Mode 2, (d) Mode 3, (e) Mode 4, and (f) Mode 6.

a high magnetizing inductance. We can assume that $i_{Lrs} = i_{Lr}/n$, and rewrite Equation (7.13) as

$$\frac{di_{Lr}}{dt} = \frac{(n-1)V_S}{n(L_{l1} + \frac{1}{n^2}L_{l2})} = \frac{(n-1)V_S}{nL_r} \qquad (7.14)$$

where L_r is the equivalent inductance of the transformer $L_{l1} + L_{l2}/n^2$. The transformer primary winding current i_{Lr} increases linearly, and the mode terminates when $i_{Lr} = I_O$. The interval of this mode can be determined by

$$\Delta t_1 = t_1 - t_0 = \frac{nL_r I_O}{(n-1)V_S} \qquad (7.15)$$

Mode 2 [shown in Figure 7.17c] $t_1 < t < t_2$: At $t = t_1$, all load current flows through the transformer primary winding, and the freewheeling diode D_3 is turned off in the ZCS condition. The freewheeling diode reverse recovery problems are greatly reduced. The snubber capacitor C_{rb} resonates with the transformer, and the voltage drop u_{S6} across switch S_6 decays. By redefining the initial time, the transformer currents i_{Lr}, i_{Lrs} and capacitor voltage u_{S6} obey the equation

$$
\begin{cases}
u_{S6}(t) = L_{l1}\dfrac{di_{Lr}(t)}{dt} + a^2 L_{l2}\dfrac{d[i_{Lrs}(t)/a]}{dt} + aV_S \\[3mm]
-C_r\dfrac{du_{S6}(t)}{dt} = i_{Lr}(t) - I_O
\end{cases}
\tag{7.16}
$$

where C_r is the capacitance of snubber capacitor C_{rb}. The transformer current $i_{Lrs} = i_{Lr}/n$, as in mode 1, with initial conditions $u_{S6}(0) = V_S$, $i_{Lr}(0) = I_O$; then the solution of (7.16) is

$$
\begin{cases}
u_{S6}(t) = \dfrac{(n-1)V_S}{n}\cos(\omega_r t) + \dfrac{V_S}{n} \\[3mm]
i_{Lrs}(t) = I_O + \dfrac{(n-1)V_S}{n}\sqrt{\dfrac{C_r}{L_r}}i_{Lr}\,\sin(\omega_r t)
\end{cases}
\tag{7.17}
$$

where $\omega_r = \sqrt{1/(L_r C_r)}$.

Let $u_{Cr}(t) = 0$, which shows the duration of the resonance

$$
\Delta t_2 = t_2 - t_1 = \frac{1}{\omega_r}\arccos\left(-\frac{1}{n-1}\right)
\tag{7.18}
$$

The interval is independent of the load current. At $t = t_2$, the corresponding transformer primary current is

$$
i_{Lr}(t_2) = I_O + V_S\sqrt{\frac{(n-2)C_r}{nL_r}}
\tag{7.19}
$$

The peak value of the transformer primary current can also be determined:

$$
i_{Lr-m} = I_O + \frac{n-1}{n}V_S\sqrt{\frac{C_r}{L_r}}
\tag{7.20}
$$

Mode 3 [shown in Figure 7.17d] $t_2 < t < t_3$: When the capacitor voltage u_{S6} reaches zero at $t = t_2$, the freewheeling diode D_{pf} begins to conduct. The current flowing through auxiliary switch S_b is the load current I_O. The sum of the currents flowing through switch S_b and diode D_{pf} is the transformer primary winding current i_{Lr}. The transformer primary voltage is zero, and the secondary voltage is V_S. By redefining the initial time, we obtain

$$0 = L_{l1}\frac{di_{Lr}(t)}{dt} + a^2 L_{l2}\frac{d[i_{Lrs}(t)/a]}{dt} + aV_S \tag{7.21}$$

Since the transformer current $i_{Lrs} = i_{Lr}/n$ as in Mode 1, we deduce Equation (7.22)

$$\frac{di_{Lr}}{dt} = -\frac{V_S}{nL_r} \tag{7.22}$$

The transformer primary current decays linearly, and the mode terminates while $i_{Lr} = I_0$. With the initial condition given by (7.19), the interval of this mode can be determined by

$$\Delta t_3 = t_3 - t_2 = \sqrt{n(n-2)L_r C_r} \tag{7.23}$$

The interval is also independent of the load current. During this mode, switch is turned on in ZVS condition.

Mode 4 [shown in Figure 7.17e] $t_3 < t < t_4$: The transformer primary winding current i_{Lr} decays linearly from load current I_O to zero. Partial load current flows through the main switch S_6. The sum of the currents flowing through switches S_6 and S_b is equal to the load current I_O. The sum of the currents flowing through switch S_b and diode D_{fp} is the transformer primary winding current i_{Lr}. By redefining the initial time, the transformer winding current obeys Equation (7.22) with the initial condition $i_{Lr}(0) = I_O$. The interval of this mode is

$$\Delta t_4 = t_4 - t_3 = \frac{nL_r I_O}{V_S} \tag{7.24}$$

The auxiliary switch S_b can be turned off in ZVS condition. In this case, after switch S_b is turned off, the transformer primary winding current flows through the freewheeling diode D_{fp}. The auxiliary switch S_b can be also be turned off in ZVS and zero current switching (ZCS) condition after i_{Lr} decays to zero.

Mode 5 [$t_4 < t < t_5$]: The transformer primary winding current decays to zero and the resonant circuit idles. This is likely the same operational state as the conventional hard-switching inverter. The load current flows from the DC link through the two switches S_1 and S_6, and the motor.

Mode 6 [shown in Figure 7.17f] $t_5 < t < t_6$: The main inverter switch S_6 is turned directly off, and the resonant circuit does not work. The snubber capacitor C_{rb}

can slow down the rise rate of u_{S6}, while the main switch S_6 operates in ZVS condition. The duration of the mode is

$$\Delta t_7 = t_7 - t_6 = \frac{C_r V_S}{I_O} \tag{7.25}$$

The next period starts from Mode 0 again, but the load current flows through freewheeling diode D_3. During phase current commutation, the switching state is changed from one lower switch to another (e.g., turn off S_6 and turn on S_2). S_6 can be turned off directly in ZVS condition (similar to Mode 6). Turning on auxiliary switch S_c to discharge the snubber capacitor C_{rc}, switch S_2 can then get ZVS condition (similar to Modes 1–4).

7.2.3 Design Considerations

It is assumed that the inductance of BDCM is much higher than the transformer leakage inductance. From the previous analysis, the design considerations can be summarized as follows:

1. Determine the value of snubber capacitor C_r, and the parameter of transformer.
2. Select the main and auxiliary switches.
3. Design the control circuitry for the main and auxiliary switches.

The turn ratio (1:n) of the transformer can be determined ahead. Equation (7.18) must satisfy

$$n > 2 \tag{7.26}$$

On the other hand, from Equation (7.24) the transformer primary winding current i_{Lr} will take a long time to decay to zero if n is too large. So n must be a relatively small number. The equivalent inductance of the transformer $L_r = L_{l1} + L_{l2}/n^2$ is inversely proportional to the rise rate of the switch current when the auxiliary switches are turned on. This means that the equivalent inductance L_r should be sufficient to limit the rising rate of the switch current to work in ZCS condition. The selection of L_r can be referred to the rule depicted in Reference [21]:

$$L_r \approx 4 t_{on} V_S / I_{O\max} \tag{7.27}$$

where t_{on} is the turn-on time of an IGBT and $I_{O\max}$ is the maximum load current. The snubber capacitance C_r is inversely proportional to the rise rate of the switch voltage drop when turning off the lower main inverter switches. It means that the capacitance should be as high as possible to limit the rising rate of the voltage to work in ZVS condition. The selection of the snubber capacitor can be determined as follows:

$$C_r \approx 4t_{on}I_{0\max}/V_S \tag{7.28}$$

where t_{off} is the turn-off time of an IGBT. However, as the capacitance increases, more energy is stored on it. This energy should be discharged when the lower main inverter switches are turned on. With high capacitance, the peak value of the transformer current will also be high. The peak value of i_{L_r} should be restricted to twice that of the maximum load current. From Equation (7.20), we obtain

$$\sqrt{\frac{C_r}{L_r}} \leq \frac{nI_{0\max}}{(n-1)V_S} \tag{7.29}$$

Three lower switches of the inverter (i.e., S_4, S_6, S_2) are turned on during Mode 3 (i.e., lag rising edge of PWM at the time range $\Delta t_1 + \Delta t_2 \sim \Delta t_1 + \Delta t_2 + \Delta t_3$). In order to turn on these switches at a fixed time (say ΔT_1), lagging rising edge of PWM under various load current can be used for convenient control. The following condition should be satisfied:

$$\Delta t_1 + \Delta t_2 + \Delta t_3 \,|_{I_0=0} > (\Delta t_1 + \Delta t_2)\,|_{I_0=I_{0\max}} + t_{off} \tag{7.30}$$

Substitute Equations (7.15), (7.18), and (7.23) into Equation (7.30):

$$\sqrt{n(n-2)L_rC_r} > \frac{nL_rI_{0\max}}{(n-1)V_S} + t_{off} \tag{7.31}$$

The whole switching transition time is expressed as

$$T_w = \Delta t_1 + \Delta t_2 + \Delta t_3 + \Delta t_4 = \frac{nL_rI_0}{(n-1)V_S} + \sqrt{L_rC_r} \times \left[\arccos\left(-\frac{1}{n-1}\right) + \sqrt{n(n-2)}\right] \tag{7.32}$$

For high switching frequencies, T_w should be as short as possible. Select the equivalent inductance L_r and snubber capacitance C_r to satisfy Equations (7.26)–(7.31), and L_r and C_r should be as small as possible.

As the transformer operates at high frequency (20 kHz), the magnetic core material can be ferrite. The design of the transformer needs the parameters of form factor, frequency, the input/output voltage, input/output maximum current, and ambient temperature. From Figure 7.16, the transformer current can be simplified as triangle waveforms, and the form factor can be then determined as $2/\sqrt{3}$. The ambient temperature is dependent on the application field. Other parameters can be obtained from the previous section. The transformer only carries current during the transition of turning on a switch in one cycle, so the winding can be a smaller diameter one.

The main switches S_{1-6} work under ZVS condition; therefore, the voltage stress is equal to the DC link voltage V_S. The device current rate can be load current. Auxiliary switches S_{a-c} work under the ZCS or ZVS condition, while the voltage stress is also equal to the DC link voltage V_S. The peak current flowing through them is limited to double the maximum load current. As the auxiliary switches S_{a-c} carry the peak current only during switch transitions, they can be rated with a lower continuous current rating. The additional cost will not be too much.

The gate signal generator circuit is shown in Figure 7.18. The rotor position signal decode module produces the typical gate signal of the main switches. The inputs of the module are rotor position signals, rotating the direction of the motor, which "enables" the signal and PWM pulse train. The rotor position signals are three square waves with a phase shift in 120°. The "enable" signal is used to disable all outputs in case of emergency (e.g., overcurrent, overvoltage, and overheat). The PWM signal is the output of comparator, comparing the reference voltage signal with the triangular wave. The reference

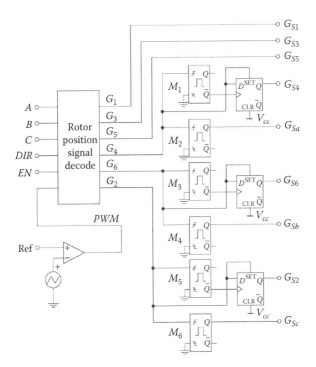

FIGURE 7.18
Gate signal generator circuit.

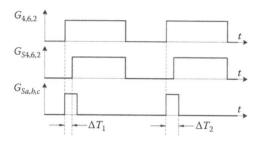

FIGURE 7.19
Gate signals $G_{S4,6,2}$ and $G_{Sa,b,c}$ from $G_{4,6,2}$.

voltage signal is the output of the speed controller. The speed controller is a processor (single-chip computer or digital signal processor), and the PWM signal can be produced by software. The outputs (G_1–G_6) of the module are the gate signals applying to the main inverter switches. The outputs $G_{1,3,5}$ are the required gate signals for the three upper main inverter switches.

The gate signals of the three lower main inverter switches and auxiliary switches can be deduced from the outputs $G_{4,6,2}$ as shown in Figure 7.19. The trailing edge of the gate signals for three lower main inverter switches $G_{S4,6,2}$ is the same as that of $G_{4,6,2}$, and the leading edge of $G_{S4,6,2}$ lags $G_{4,6,2}$ for a short time ΔT_1. The gate signals for auxiliary switches $G_{S4,6,2}$ have a fixed pulse width (ΔT_2), the leading edge being the same as that of $G_{4,6,2}$. The gate signals $G_{Sa,b,c}$ are the outputs of monostable flip-flops $M_{4,6,2}$ with the inputs $G_{4,6,2}$. The three monostable flip-flops $M_{4,6,2}$ have the same pulse width ΔT_2. The gate signals $G_{S4,6,2}$ are combined by the negative outputs of monostable flip-flops $M_{1,3,5}$ and $G_{4,6,2}$. The combining logical controller can be implemented by a D flip-flop with "preset" and "clear" terminals. The three monostable flip-flops $M_{4,6,2}$ have the same pulse width ΔT_1. Determination of the pulse widths of ΔT_1 and ΔT_2 is from theoretical analysis in the preceding subsection. In order to get the ZVS condition of the main inverter switches under various load currents, the lag time should satisfy

$$(\Delta t_1 + \Delta t_2)\big|_{I_O = I_{O\,max}} < \Delta T_1 < (\Delta t_1 + \Delta t_2 + \Delta t_3)\big|_{I_O = 0} - t_{off} \tag{7.33}$$

In order to get a soft-switching condition of the auxiliary switches, pulse width need only satisfy

$$\Delta T_2 > (\Delta t_1 + \Delta t_2 + \Delta t_3)\big|_{I_O = I_{O\,max}} \tag{7.34}$$

7.2.4 Simulation and Experimental Results

The proposed topology is verified by the simulation software PSim. The DC link voltage is 300 V, and the maximum load current is 25 A. The parameters

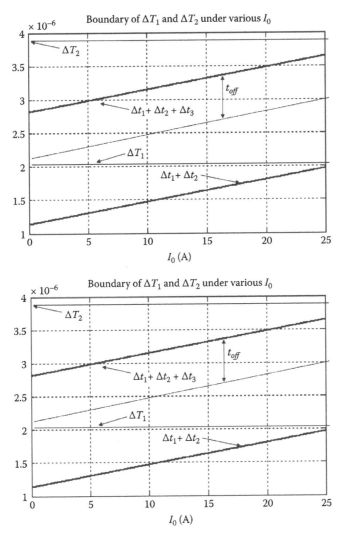

FIGURE 7.20
Boundary of ΔT_1 and ΔT_2 under various load currents I_O.

of the resonant circuit were determined from Equations (7.26)–(7.32). The transformer turns ratio is 1:4, and the leakage inductances of the primary secondary windings are 6 and 24 μH, respectively. Therefore, the equivalent transformer inductance L_r is 7.5 μH. The resonant capacitance C_r is 0.047 μF. Then, $\Delta t_1 + \Delta t_2$ and $\Delta t_1 + \Delta t_2 + \Delta t_3$ can be determined under various load currents I_O as shown in Figure 7.20, considering that the turn-off times of a switch with lagging time ΔT_1 and pulse width ΔT_2 are set to 2.1 μs and 5 μs, respectively. The frequency of the PWM is 20 kHz. Waveforms of transformer

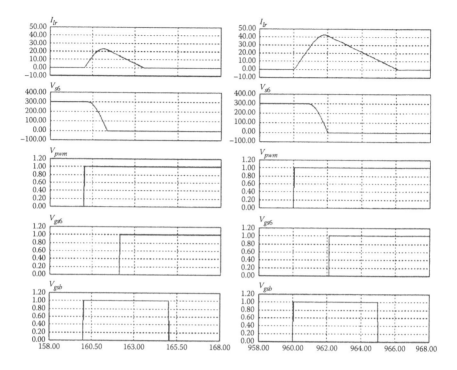

FIGURE 7.21

Simulation waveforms of i_{Lr}, u_{S6}, PWM, S_6, and S_b gate signal under various load currents: (a) under low load current ($I_O = 5$ A) and (b) under high load current ($I_O = 25$ A).

primary winding current i_{Lr}, switch S_6 voltage drop u_{S6}, PWM, main switch S_6, and auxiliary switch S_b, and the gate signal under low and high load currents are shown in Figure 7.21. The figure shows that the inverter worked well under various load currents. In order to verify the theoretical analysis and simulation results, the inverter was tested by experiment. The test conditions are

1. DC link voltage: 300 V
2. Power of BDCM: 3.3 Hp
3. Rated phase current: 7.8 A
4. Switching frequency: 20 kHz

Select a 50 A, 1200 V BSM 35 GB 120 DN2 dual IGBT module as main inverter switches, and a 30 A 600 V IMBH30D-060 IGBT as auxiliary switches. With datasheets of these switches and Equations (7.26)–(7.32), the value of inductance and capacitance can be determined. Three polyester capacitors of 47 nF/630 V were adopted as snubber capacitors for the three lower switches of the inverter. A highly magnetizing inductance transformer with turn ratio 1:4 was employed in the experiment. Fifty-two turn wires of size AWG 15 are selected as the primary winding, and 208 turn wires with size AWG 20 are selected as the secondary winding. The equivalent inductance is about 7 μH. The switching frequency is 20 kHz. The rotor position signal decode module is implemented by a 20 lead gate array logic (GAL) IC GAL16V8. The monostable flip-flop is set up by IC 74LS123, variable resistor, and capacitor. With Equations (7.33) and (7.34), lag time and pulse width are determined to be 2.5 and 5 μs, respectively.

The system is tested in light load and full load currents. The voltage waveforms across the main inverter switch u_{S6} and its gate signal with low and high load currents are shown in Figures 7.22a and 7.22b, respectively. All the voltage signals are measured by a differential probe with a gain of 20, for voltage waveform, 5.00 V/div 100 V/div. The waveforms of u_{S6} and the current i_{S6} are shown in Figure 7.22c. We can see that both dv/dt and di/dt are reduced

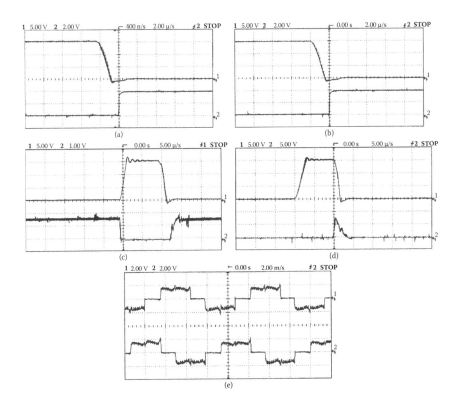

FIGURE 7.22
Experiment waveforms: (a) Switch S_6 voltage u_{S6} (top) and its gate signal (bottom) under low load current (100 V/div), (b) Switch S_6 voltage u_{S6} (top) and its gate signal (bottom) under high load current (100 V/div), (c) Switch S_6 voltage u_{S6} (top) and its current i_{S6} (bottom) (100 V/div, 5 A/div), (d) Switch S_6 voltage u_{S6} (top) and transformer current i_{S6} (bottom) (100 V/div, 25 A/div), (e) waveforms of phase current (10 A/div).

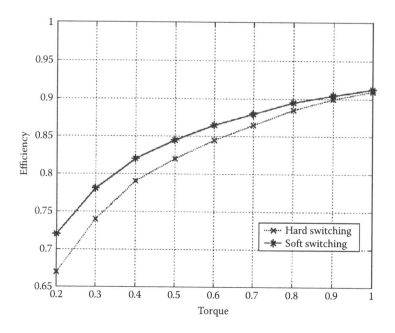

FIGURE 7.23
Efficiency of hard switching and soft switching under various load torques (p.u.).

significantly. The waveforms of u_{S6} and transformer primary winding current i_{Lr} are shown in Figure 7.22d. The phase current is shown in Figure 7.22e. It can be seen that the resonant pole inverter works well under various load currents, and there is little overlap between the voltage and current waveforms during soft-switching condition; therefore, the switching. The efficiency of hard switching and soft switching under rated speeds and various load torques (p.u.) are shown in Figure 7.23. Efficiency improves with the soft-switching inverter. Therefore, the design of the system is successful.

7.3 Transformer-Based Resonant DC Link Inverter

In order to generate voltage notches of the DC link at controllable instants and reduce the power losses of the inductor, several quasi-parallel resonant schemes were proposed [5,6,22]. As a dwell time is generally required after every notch, severe interference occurs, mainly in multiphase inverters, appreciably worsening the modulation quality. A novel DC rail parallel resonant zero voltage transition (ZVT) voltage source inverter [23] was introduced that overcomes many drawbacks mentioned earlier. However, it

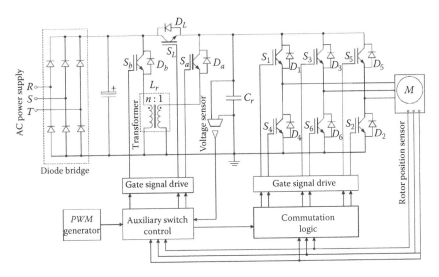

FIGURE 7.24
Structure of the resonant DC link inverter for the BDCM drive system.

requires a stiff DC link capacitor bank that is center-taped to accomplish commutation. The center voltage of the DC link is susceptible to drift, which may affect the operation of the resonant circuit. In addition, it requires two ZVT per PWM cycle, which would lower the output voltage and limit the switch frequency of the inverter.

On the other hand, the majority of soft-switching inverters proposed in recent years have been aimed at induction motor drive applications. So it was necessary to research the novel topology of soft-switching inverter and the special control circuit for BDCM drive systems. This chapter proposes a resonant DC link inverter based on a transformer for the BDCM drive system to solve the aforementioned problems. The inverter possesses the advantages of low switching power loss, low inductor power loss, low DC link voltage ripple, small device voltage stress, and a simple control scheme. The structure of the soft-switching inverter is shown in Figure 7.24 [24]. The system contains a diode bridge rectifier, a resonant circuit, a conventional three-phase inverter, and a control circuit. The resonant circuit consists of three auxiliary switches (S_L, S_a, S_b) and corresponding built-in freewheeling diode (D_L, D_a, D_b), one transformer with turn ratio 1:n, and one resonant capacitor. All auxiliary switches work under ZVS or zero current switching (ZCS) condition. The system generates voltage notches of the DC link to guarantee that the main switches (S_1–S_6) of the inverter operate in ZVS condition.

7.3.1 Resonant Circuit

The resonant circuit consists of three auxiliary switches, one transformer, and one resonant capacitor. The auxiliary switches are controlled at a

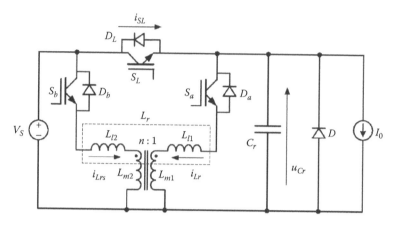

FIGURE 7.25
Equivalent circuit of the inverter.

certain instant to obtain resonance between transformer and capacitor. Thus, the DC link voltage reaches zero temporarily (voltage notch), and the main switches of the inverter reach ZVS condition for commutation. Since the resonant process is very short, the load current can be assumed constant. The equivalent circuit of the inverter is shown in Figure 7.25. When V_S is the DC power supply voltage, I_O is the load current. The corresponding waveforms of the auxiliary switches gate signal, PWM signal, resonant capacitor voltage u_{Cr} (i.e., DC link voltage), the transformer primary winding current i_{Lr}, and current i_{SL} of switch (S_L) are illustrated in Figure 7.26. The DC link voltage is reduced to zero and then rises to the supply voltage again; this is called one zero voltage transition (ZVT) process or one DC link voltage notch. The operation of the ZVT process in one PWM cycle can be divided into eight modes.

Mode 0 [shown in Figure 7.27a] $0 < t < t_0$: Its operation is the same as the conventional inverter. Current flows from DC power supply through S_L to the load. The voltage u_{Cr} across resonant capacitor C_r is equal to the supply voltage V_S. The auxiliary switches S_a and S_b are turned off.

Mode 1 [shown in Figure 7.27b] $t_0 < t < t_1$: When it is the instant for phase current commutation or the PWM signal is flipped from high to low, the auxiliary switch S_a is turned on with ZCS (as the i_{Lr} cannot suddenly change due to the transformer inductance) and switch S_L is turned off with ZVS (as the voltage cannot be suddenly changed due to resonant capacitor C_r) at the same time. The transformer primary winding current i_{Lr} begins to increase, and the secondary winding current i_{Lrs} also begins to build up through diode D_b to the DC link. The terminal voltages of primary and secondary windings of the transformer are the DC link voltage u_{Cr} and supply voltage V_S,

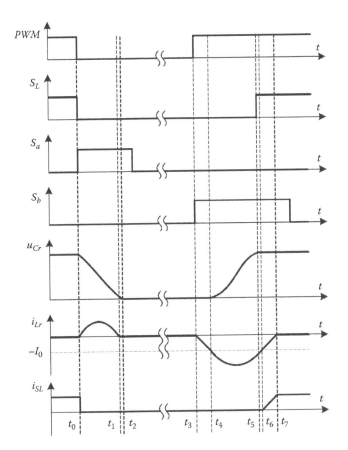

FIGURE 7.26
Key waveforms of the equivalent circuit.

respectively. Capacitor C_r resonates with the transformer, and the DC link voltage u_{Cr} is decreased. Neglecting the resistances of windings and using the transformer equivalent circuit (referred to the primary side) [25], the transformer current i_{Lr}, i_{Lrs}, and DC link voltage u_{Cr} obey the equation

$$\begin{cases} u_{Cr}(t) = L_{l1}\dfrac{di_{Lr}(t)}{dt} + a^2 L_{l2}\dfrac{d[i_{Lrs}(t)/a]}{dt} + aV_s \\[2mm] i_{Lr}(t) + I_O + C_r\dfrac{du_{Cr}(t)}{dt} = 0 \end{cases} \tag{7.35}$$

where L_{l1} and L_{l2} are the primary and secondary winding leakage inductance, respectively. The transformer turn ratio is 1:n. The transformer has

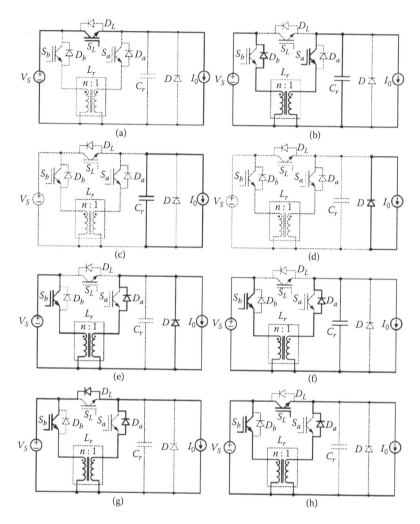

FIGURE 7.27
Operation mode of the resonant DC link inverter: (a) Mode 0, (b) Mode 1, (c) Mode 2, (d) Mode 3, (e) Mode 4, (f) Mode 5, (g) Mode 6, and (h) Mode 7.

a high magnetizing inductance. We can assume that $i_{Lrs} = i_{Lr}/n$, with initial condition $u_{Cr}(0) = V_S$, $i_{Lr}(0) = 0$; solving Equation (7.35), we get

$$
\begin{cases}
u_{Cr}(t) = \dfrac{(n-1)V_S}{n}\cos(\omega_r t) - I_O\sqrt{\dfrac{L_r}{C_r}}\sin(\omega_r t) + \dfrac{V_S}{n} \\[3mm]
i_{Lr}(t) = I_O\cos(\omega_r t) - I_O + \dfrac{(n-1)V_S}{n}\sqrt{\dfrac{L_r}{C_r}}\sin(\omega_r t)
\end{cases}
\tag{7.36}
$$

where $L_r = L_{l1} + L_{l2}/n^2$ is the equivalent inductance of the transformer, and $\omega_r = \sqrt{(1/L_rC_r)}$ is the natural angular resonance frequency. Rewriting Equation (7.36), we get

$$
\left\{
\begin{aligned}
u_{Cr}(t) &= K\cos(\omega_r t + \alpha) + \frac{V_S}{n} \\
i_{Lr}(t) &= K\sqrt{\frac{C_r}{L_r}}\sin(\omega_r t + \alpha) - I_O
\end{aligned}
\right.
\tag{7.37}
$$

where $K = \sqrt{((n-1)^2 V_S^2/n^2 + (I_0^2 L_r/C_r))}$, $\alpha = \arctan[(nI_0\sqrt{L_r/C_r}/(n-1)V_S]$. n is slightly less than 2 (the selection of the number will be explained later), and i_{Lr} will decay to zero faster than u_{Cr}. Letting $i_{Lr}(t) = 0$, the duration of the resonance can be determined:

$$
\Delta t_1 = t_1 - t_0 = \frac{\pi - \alpha}{\omega_r}
\tag{7.38}
$$

When i_{Lr} is reduced to zero, the auxiliary switch S_a can be turned off with ZCS condition. At $t = t_1$, the corresponding dc link voltage u_{Cr} is

$$
u_{Cr}(t_1) = \frac{2-n}{n}V_S
\tag{7.39}
$$

Mode 2 [shown in Figure 7.27c] $t_1 < t < t_2$: When the transformer current is reduced to zero, the resonant capacitor is discharged through load from initial condition as in Equation (7.39). The interval of this mode can be determined by

$$
\Delta t_2 = t_2 - t_1 = \frac{C_r V_S(2-n)}{nI_0}
\tag{7.40}
$$

As has been mentioned, n is slightly less than 2, and the interval is normally very short.

Mode 3 [shown in Figure 7.27d] $t_2 < t < t_3$: The DC link voltage u_{Cr} is zero. The main switches of the inverter can now be either turned on or turned off under ZVS condition during this mode. Load current flows through the freewheeling diode D.

Mode 4 [shown in Figure 7.27e] $t_3 < t < t_4$: As the main switches have turned on or turned off, the auxiliary switch S_b is turned on with ZCS (as i_{Lrs} cannot suddenly change due to the transformer inductance) and the transformer secondary current i_{Lrs} starts to build up linearly. The transformer primary current i_{Lr} also begins to conduct through diode D_a to the load. The current in the freewheeling diode D begins to fall linearly. The load current is slowly diverted from the freewheeling diodes to the resonant circuit. The DC link voltage u_{Cr} is still zero before the transformer primary current is greater than load current. The terminal voltages of transformer primary and secondary windings are zero and DC power supply voltage V_S, respectively. Redefining the initial time, we obtain

$$0 = L_{l1} \frac{di_{Lr}(t)}{dt} + a^2 L_{l2} \frac{d[i_{Lrs}(t)/a]}{dt} + aV_S \tag{7.41}$$

Since the transformer current $i_{Lrs} = i_{Lr}/n$ as in mode 1, we rewrite Equation (7.41) as

$$\frac{di_{Lr}}{dt} = -\frac{V_S}{nL_r} \tag{7.42}$$

The transformer primary current is increased reverse-linearly from zero, and the mode terminates when $i_{Lr} = -I_O$; the interval of this mode can be determined as

$$\Delta t_4 = t_4 - t_3 = \frac{nL_r I_O}{V_S} \tag{7.43}$$

At t_4, i_{Lr} equals the negative load current $-I_O$, and the current through the diode D becomes zero. Thus, the freewheeling diode turns off under ZCS condition, and the diode reverse recovery problems are reduced.

Mode 5 [shown in Figure 7.27f] $t_4 < t < t_5$: The absolute value of i_{Lr} is continuously increased from I_O, and u_{Cr} is increased from zero when the freewheeling diode D is turned off. Redefining the initial time, we can get the same equation as Equation (7.35). The initial condition is $u_{Cr}(0) = 0$, $i_{Lr}(0) = -I_O$; neglecting the inductor resistance and solving the equation, we get

$$\begin{cases} u_{Cr}(t) = -\frac{V_S}{n}\cos(\omega_r t) + \frac{V_S}{n} \\ \\ i_{Lr}(t) = -I_O - \frac{V_S}{n}\sqrt{\frac{C_r}{L_r}}\sin(\omega_r t) \end{cases} \tag{7.44}$$

when

$$\Delta t_5 = t_5 - t_4 = \frac{1}{\omega_r}\arccos(1 - n) \tag{7.45}$$

$u_{Cr} = V_S$, and the auxiliary switch S_L is turned on with ZVS (due to C_r). The interval is independent of load current. At $t = t_5$, the corresponding transformer primary current i_{Lr} is

$$i_{Lr}(t_5) = -I_O - V_S\sqrt{\frac{(2-n)C_r}{nL_r}} \qquad (7.46)$$

The peak value of the transformer primary current can also be determined:

$$i_{Lr-m} = \left| -I_O - \frac{V_S}{n}\sqrt{\frac{C_r}{L_r}} \right| = I_O + \frac{V_S}{n}\sqrt{\frac{C_r}{L_r}} \qquad (7.47)$$

Mode 6 [shown in Figure 7.27g] $t_5 < t < t_6$: Both the terminal voltages of primary and secondary windings are equal to the supply voltage V_S after the auxiliary switch S_L is turned on. Redefining the initial time, we obtain

$$V_S = L_{l1}\frac{di_{Lr}(t)}{dt} + a^2 L_{l2}\frac{d[i_{Lrs}(t)/a]}{dt} + aV_S \qquad (7.48)$$

Since the transformer current $i_{Lrs} = i_{Lr}/n$ as in mode 1, we rewrite Equation (7.48) as

$$\frac{di_{Lr}}{dt} = \frac{(n-1)V_S}{nL_r} \qquad (7.49)$$

The transformer primary current i_{Lr} decays linearly, and the mode terminates when $i_{Lr} = -I_O$ again. With the initial condition (Equation 7.46), the interval of this mode can be determined as

$$\Delta t_6 = t_6 - t_5 = \frac{\sqrt{n(2-n)L_r C_r}}{n-1} \qquad (7.50)$$

The interval is also independent of load current. As mentioned earlier, n is slightly less than 2, and the interval is also very short.

Mode 7 [shown in Figure 7.27h] $t_6 < t < t_7$: The transformer primary winding current i_{Lr} decays linearly from negative load current $-I_O$ to zero. Partial load current flows through the switch S_L. The sum of the currents flowing through switch S_L and transformer is equal to the load current I_O. Redefining the initial time, the transformer winding current obeys Equation (7.49) with the initial condition $i_{Lr}(0) = -I_O$. The interval of this mode is

$$\Delta t_7 = t_7 - t_6 = \frac{nL_r I_O}{(n-1)V_S} \qquad (7.51)$$

Then auxiliary switch S_b can be also turned off with ZCS condition after i_{Lr} decays to zero (at any time after t_7).

7.3.2 Design Considerations

It is assumed that the inductance of BDCM is much higher than the transformer leakage inductance. From the analysis presented previously, the design considerations can be summarized as follows:

1. Determine the value of resonant capacitor C_r and the parameters of the transformer.
2. Select the main switches and auxiliary switches.
3. Design the gate signal for the auxiliary switches.

The turn ratio 1:n of the transformer can be determined ahead. From Equation (7.45), n must satisfy

$$n < 2 \tag{7.52}$$

On the other hand, from Equations (7.39) and (7.40), n should be as close to 2 as possible so that the duration of mode 2 would be not very long and would be small enough at the end of mode 1.

Normally, n can be selected in the range 1.7–1.9. The equivalent inductance of the transformer $L_r = L_{l1} + L_{l2}/n^2$ is inversely proportional to the rising rate of switch current when auxiliary switches are turned on. It means that the equivalent inductance L_r should be large enough to limit the rising rate of the switch current to work in ZCS condition. The selection of L_r can be referred to the rule depicted in Reference [26].

$$L_r \geq \frac{4t_{on}V_S}{I_{O\max}} \tag{7.53}$$

where t_{on} is the turn-on time of switch S_a, and $I_{O\max}$ is the maximum load current. The resonant capacitance C_r is inversely proportional to the rising rate of switch voltage drop when switch S_L is turned off. This means that the capacitance is as high as possible to limit the rising rate of the voltage to work in ZVS condition. The selection of the resonant capacitor can be determined as

$$C_r \geq \frac{4t_{off}I_{O\max}}{V_S} \tag{7.54}$$

where t_{off} is the turn-off time of switch S_L. However, as the capacitance increases, more energy is stored in it, and the peak value of transformer current will also be high. The peak value of i_{Lr} should be limited to twice the peak load current. From Equation (7.47), we obtain

$$\sqrt{\frac{C_r}{L_r}} \le \frac{nI_{O\max}}{V_S} \tag{7.55}$$

The DC link voltage rising transition time is expressed as

$$T_w = \Delta t_4 + \Delta t_5 = \frac{nL_r I_{O\max}}{V_S} + \sqrt{L_r C_r} \arccos(1-n) \tag{7.56}$$

For high switching frequency, T_w should be as short as possible. Select the equivalent inductance L_r and resonant capacitance C_r to satisfy the inequalities (7.52)–(7.55); L_r and C_r should be as small as possible. L_r and C_r selection area is illustrated in Figure 7.28 to determine their values, and the valid area

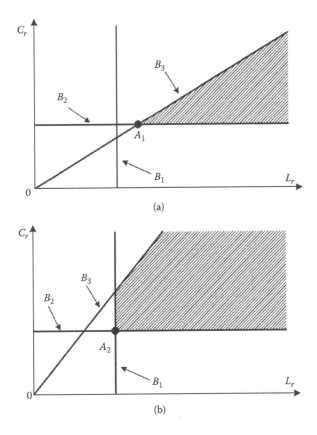

FIGURE 7.28
L and C selection area: (a) Case 1: B intersects B first and (b) Case 2: B intersects A first.

is shadowed, where B_1–B_3 is the boundary, which is defined according to inequalities (7.52)–(7.55):

$$B_1 : L_r = \frac{4t_{on}V_S}{I_{0\max}} \tag{7.57}$$

$$B_2 : C_r = \frac{4t_{off}I_{0\max}}{V_S} \tag{7.58}$$

$$B_3 : \sqrt{\frac{C_r}{L_r}} = \frac{nI_{0\max}}{V_S} \tag{7.59}$$

If boundary B_3 intersects B_1 first as shown in Figure 7.28a, the value of L_r and C_r in the intersection (i.e., A_1) can be selected. Otherwise, the value of L_r and C_r in the intersection A_2 is selected as shown in Figure 7.28b.

Main switches S_1–S_6 work under ZVS condition, and the voltage stress is equal to the DC power supply voltage V_S. The device current rate can be the load current. The auxiliary switch S_L works under ZVS condition, and its voltage and current stresses are the same as for the main switches. Auxiliary switches S_a and S_b work under ZCS or ZVS condition, and the voltage stress is equal to the DC power supply voltage V_S. The peak current flowing through them is limited to double the maximum load current. As the auxiliary switches S_a and S_b carry the peak current only during switch transitions, they can be the devices with lower continuous current ratings.

The design of the gate signal for auxiliary switches can be referenced from Figure 7.26. The trailing edge of the gate signal for auxiliary switch S_L is the same as that of PWM, and the leading edge is determined by the output of the DC link voltage sensor. The gate signal for auxiliary switch S_a is a positive pulse with leading edge the same as the PWM trailing edge, and its width ΔT_a should be greater than Δt_1. From Equation (7.38), Δt_1 is maximum when the load current is zero. So ΔT_a can be a fixed value determined by

$$\Delta T_a > \Delta t_1 |_{\max} = \frac{\pi}{\omega_r} = \pi\sqrt{L_r C_r} \tag{7.60}$$

The gate signal for auxiliary switch S_b is also a pulse with leading edge the same as that of PWM, and its width ΔT_b should be longer than $t_7 - t_3$ (i.e., $\Delta t_4 + \Delta t_5 + \Delta t_6 + \Delta t_7$). ΔT_b can be determined from Equations (7.43), (7.45), (7.50), and (7.51):

$$\Delta T_b > \sum_{i=4}^{7} \Delta t_i |_{\max} = \frac{n^2 L_r I_{0\max}}{(n-1)V_S} + \sqrt{L_r C_r} \times \left[\arccos(1-n) + \frac{\sqrt{n(2-n)}}{n-1} \right] \tag{7.61}$$

7.3.3 Control Scheme

When the duty of PWM is 100%, that is, full duty cycle, the main switches of the inverter work under the commutation frequency. When it is the instant to commutate the phase current of the BDCM, we control the auxiliary switches S_a, S_b, and S_L, and resonance occurs between transformer inductor L_r and capacitor C_r. The DC link voltage reaches zero temporarily; thus, ZVS condition of the main switches is obtained. When the duty of PWM is less than 100%, the auxiliary switch S_L works as a chopper. The main switches of the inverter do not switch within a PWM cycle when the phase current need not commutate. It has the benefit of reducing phase current drop when the PWM is off. The phase current is commutated when the DC link voltage becomes zero. There is only one DC link voltage notch per PWM cycle. It is very important especially for very low or very high duty of PWM. Otherwise the interval between two voltage notches will be very short even when overlapped, which will limit the tuning range.

The commutation logical circuit of the system is shown in Figure 7.29. It is similar to the conventional BDCM commutation logical circuit except that six D flip-flops are added to the output. Thus, the gate signal of the main switches is controlled by the synchronous pulse CK, which will be mentioned later, and the commutation can be synchronized with the auxiliary

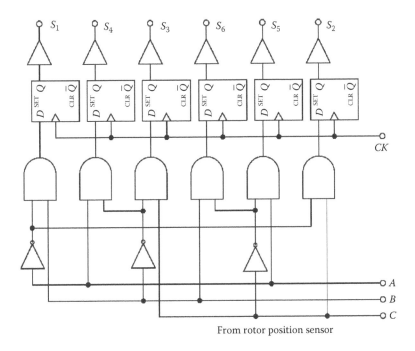

FIGURE 7.29
Commutation logical circuit for the main switches.

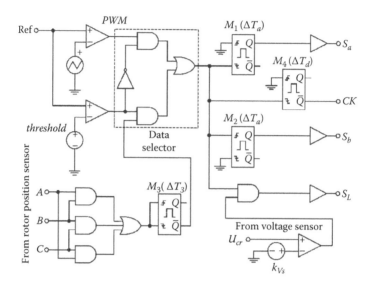

FIGURE 7.30
Control circuit for the auxiliary switches.

switches' control circuit (shown in Figure 7.30). The operation of the inverter can be divided into PWM operation and full duty cycle operation.

7.3.3.1 Full Duty Cycle Operation

When the duty of PWM is 100%, that is, full duty cycle, the whole ZVT process (mode 1–mode 7) occurs when the phase current commutation is under way. The monostable flip-flop M_3 will generate one narrow negative pulse. The width of the pulse ΔT_3 is determined by $(\Delta t_1 + \Delta t_2 + T'_c)$, where T'_c is a constant, considering the turn-on/turn-off time of the main switches. If n is close to 2, Δt_2 would be very short or u_{Cr} would be small enough at the end of the mode1, and ΔT_3 can be determined by

$$\Delta T_3 = \Delta t_1 \mid_{max} + T_C = \pi\sqrt{L_r C_r} + T_C \qquad (7.62)$$

where T_c is a constant that is greater than T'_c. The data selector makes the output of monostable flip-flop M_3 active. The monostable flip-flop M_1 generates a positive pulse when the trailing edge of M_3 negative pulse arrives. The pulse is the gate signal for auxiliary switch S_a, and its width is ΔT_a, which is determined by inequality (7.60). The gate signal for switch S_L is flipped to low at the same time. Then mode 1 begins, and the DC link voltage is reduced to zero. Synchronous pulse CK is also generated by a monostable flip-flop M_4, and the pulse width ΔT_d should be greater than maximum Δt_1 (i.e., $\pi\sqrt{L_r C_r}$). If the D flip-flops are rising edge active, then CK is connected to the negative output of M_4, otherwise it is connected to the positive output. Thus, the active edge of pulse CK is within mode 3 when the voltage of the DC link is zero, and the main switches of the inverter are in ZVS condition. The monostable flip-flop M_2 generates a positive pulse when the leading edge of negative pulse arrives. The pulse width of M_2 is ΔT_d ,which is determined by inequality (7.61). Then modes 4–7 occur, and the DC link voltage is increased to that of the supply voltage again. The leading edge of the gate signal for switch S_L is determined by the DC link voltage sensor signal. In other words, in full-cycle operation, when the phase current commutation is under way, the resonant circuit generates a DC link voltage notch to let main switches of the inverter switch under ZVS condition.

7.3.3.2 PWM Operation

In this operation, the data selector makes the PWM signal active. The auxiliary switch S_L works as a chopper, but the main switches of the inverter do not turn on or turn off within a single PWM cycle when the phase current need not commutate. The load current is commutated when the DC link voltage becomes zero. (As the PWM cycle is very short, it does not affect the operation of the motor).

1. When PWM signal is flipped down, mode 1 begins, the pulse signal for switch S_a is generated by M_1, and the gate signal for switch S_L drops to low. However, the voltage of the DC link does not increase until the PWM signal is flipped up. Pulse CK is also generated by M_4 to locate the active edge of CK in mode 3.

2. When the PWM signal is flipped up, mode 4 begins, and the pulse signal for switch S_b is generated at that moment. Then, when the voltage of the DC link is increased to the supply voltage V_S, the gate signal for switch S_L is flipped to a high level.

Thus, only one ZVT occurs per PWM cycle: modes 1 and 2 for PWM turning off, and modes 4, 5, 6, and 7 for PWM turning on. And the switching frequency would not be greater than the PWM frequency.

7.3.4 Simulation and Experimental Results

The proposed system is verified by simulation software PSim. The DC power supply voltage V_S is 240 V, and the maximum load current is 12 A. The transformer turn ratio n is 1: 1.8, and the leakage inductances of the primary secondary windings are selected as 4 and 12.96 μH, respectively. So the equivalent transformer inductance L_r is about 8 μH. The resonant capacitance C_r is 0.1 μF. Switch $S_{a,b}$ gate signal widths ΔT_a and ΔT_b are set to be 3 and 6 μs, respectively. The narrow negative pulse width ΔT_3 in a full-duty cycle is set to 4.5 µs, and the delay time for synchronous pulse CK is set to 3.5 µs. The frequency of the PWM is 20 kHz. Waveforms of dc link voltage u_{Cr}, transformer primary winding current i_{Lr}, switch S_L and diode D_L current i_{SL}/i_{DL}, PWM, auxiliary switch gate signal under low and high load current are shown in Figure 7.31. The figure shows that the inverter worked well under various load currents.

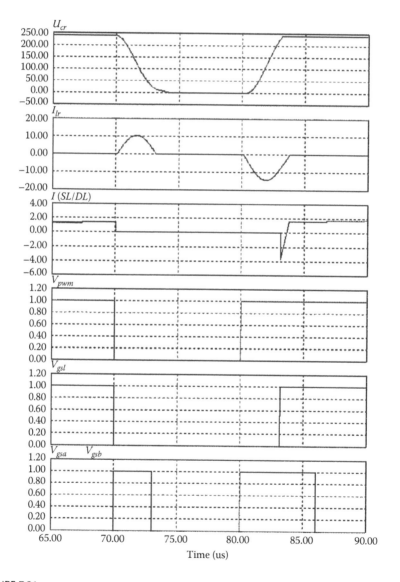

FIGURE 7.31

Waveforms of u_{Cr}, i_{Lr}, i_{SL}/i_{DL}, PWM, auxiliary switches' gate signal under various load currents: (a) under low load current ($I_O = 2$ A) and (b) under high load current ($I_O = 8$ A).

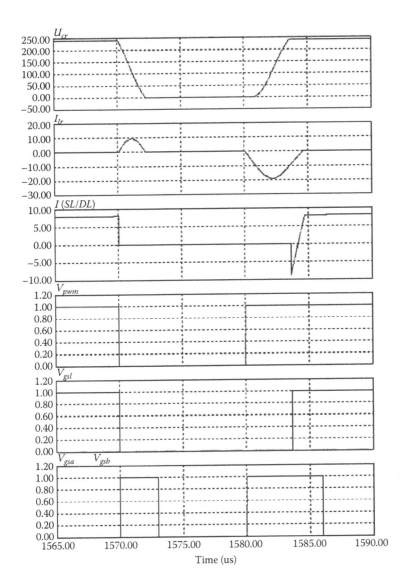

FIGURE 7.31 (continued)
Waveforms of u_{Cr}, i_{Lr}, i_{SL}/i_{DL}, PWM, auxiliary switches' gate signal under various load currents: (a) under low load current ($I_O = 2$ A) and (b) under high load current ($I_O = 8$ A).

In order to verify the theoretical analysis and simulation results, the proposed soft switching inverter was tested on an experimental prototype. The DC link voltage is 240 V, the rated phase current is 7.8 A, and the switching frequency is 20 kHz. Select a 50 A/1200 V BSM 35 GB 120 DN2 dual IGBT module as the main inverter switches S_1–S_6 and auxiliary switch S_L, another switch in the same module of S_L can be adopted as auxiliary switch S_a, and

30 A/600 V IMBH30D-060 IGBT as auxiliary switch S_b. With datasheets of these switches and Equations (7.52)–(7.55), the value of capacitance and the parameter of the transformer can be determined. A polyester capacitor of 0.1 µF, 1000 V was adopted as the DC link resonant capacitor C_r. A high magnetizing inductance transformer with turn ratio 1:1.8 was employed in the experiment. The equivalent inductance is about 8 µH under short-circuit test [25]. The switching frequency is 20 kHz. The monostable flip-flop is set up by IC 74LS123, variable resistor, and capacitor. The logical gate can be replaced by a programmable logical device to reduce the number of ICs. ΔT_a, ΔT_b, ΔT_3, and ΔT_d are set to 3, 6, 4.5, and 3.5 µs, respectively.

The system is tested under light and heavy loads. The waveform's DC link voltage u_{Cr} and transformer primary winding current i_{Lr} under low and high load currents are shown in Figures 7.32a and 7.32b, respectively. The transformer-based resonant DC link inverter works well under various load currents. The waveforms of auxiliary switch S_L voltage u_{SL} and its current i_{SL}

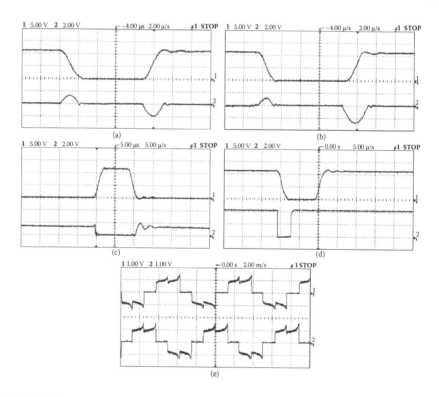

FIGURE 7.32
Experiment waveforms: (a) the DC link voltage u_{Cr} (top) and transformer current i_{Lr} (bottom) under low load current (100 V/div, 10 A/div), (b) the DC link voltage u_{Cr} (top) and transformer current i_{Lr} (bottom) under high load current (100 V/div, 10 A/div), (c) switch S_L voltage (top) and current (bottom) (100 V/div, 10 A/div), (d) the DC link voltage u_{Cr} (top) and synchronous signal CK (bottom) (100 V/div), and (e) phase current of BDCM (5 A/div).

are shown in Figure 7.32c. There is little overlap between the switch S_L voltage and its current during the switching under soft-switching condition, so the switching power losses are low. The waveforms of the resonant DC link voltage u_{Cr} and synchronous signal CK are shown in Figure 7.32d, which the main switches can switch under ZVS condition during commutation. The phase current of BDCM is shown in Figure 7.32e. The design of the system is successful.

References

1. Pan, Z. Y. and Luo, F. L. 2004. Novel soft-switching inverter for brushless DC motor variable speed drive system. *IEEE Trans. Power Electron.*, pp. 280–288.
2. Divan, D. M. 1989. The resonant dc link converter—a new concept in static power conversion. *IEEE Trans. Ind. Applicat.*, pp. 317–325.
3. Divan, D. M. and Skibinski, G. 1989. Zero-switching-loss inverters for high-power applications. *IEEE Trans. Ind. Applicat.*, pp. 634–643.
4. Yi, W., Liu, H. L., Jung, Y. C., Cho, J. G., and Cho, G. H. 1992. Program-controlled soft switching PRDCL inverter with new space vector PWM algorithm. *Proc. IEEE PESC'92*, pp. 313–319.
5. Malesani, L., Tenti, P., Tomasin, P., and Toigo, V. 1995. High efficiency quasiresonant dc link three-phase power inverter for full-range PWM. *IEEE Trans. Ind. Applicat.*, pp. 141–148.
6. Jung, Y. C., Liu, H. L., Cho, G. C., and Cho, G. H. 1995. Soft switching space vector PWM inverter using a new quasiparallel resonant dc link. *Proc. IEEE PESC*, pp. 936–942.
7. Zhengfeng, M. and Yanru, Z. 2001. A novel dc-rail parallel resonant ZVT VSI for three-phases AC motor drive. *Proc. Int. Conf. Elect. Machines Syst. (ICEMS'201)*, pp. 492–495.
8. Murai, Y., Kawase, Y., Ohashi, K., Nagatake, K., and Okuyama, K. 1989. Torque ripple improvement for brushless dc miniature motors. *IEEE Trans. Ind. Applicat.*, pp. 441–450.
9. Chang-heeWon, C., Joong-ho Song, J., and Choy, I. 2002. Commutation torque ripple reduction in brushless dc motor drives using a single dc current sensor. *Proc. IEEE PESC*, pp. 985–990.
10. Sebastian, T. and Gangla, V. 1996. Analysis of induced EMF waveforms and torque ripple in a brushless permanent magnet machine. *IEEE Trans. Ind. Applicat.*, pp. 195–200.
11. Pillay, P. P. and Krishnan, R. 1988. Modeling of permanent magnet motor drives. *IEEE Trans. Ind. Electron.*, pp. 537–541.
12. Pan, Z. Y. and Luo, F. L. 2005. Novel resonant pole inverter for brushless DC motor drive system. *IEEE Trans. Power Electron.*, pp. 173–181.
13. De Doncker, R. W. and Lyons, J. P. 1990. The auxiliary resonant commutated pole converter. *Proc. IEEE Industry Applications Soc. Annu. Meeting*, pp. 1228–1235.
14. McMurray, W. 1989. Resonant snubbers with auxiliary switches. *Proc. IEEE Industry Applications Soc. Annu. Meeting*, pp. 289–834.

15. Vlatkovic, V., Borojevic, D., Lee, F., Cuadros, C., and Gataric, S. 1993. A new zero-voltage transition, three-phase PWM rectifier/inverter circuit. *Proc. IEEE PESC*, pp. 868–873.
16. Cuadros, C., Borojevic, D., Gataric, S., and Vlatkovic, V. 1994. Space vector modulated, zero-voltage transition three-phase to DC bidirectional converter. *Proc. IEEE PESC*, pp. 16–23.
17. Lai, J. S., Young Sr., R.W., Ott Jr., G.W., White, C. P., McKeever, J.W., and Chen, D. 1995. A novel resonant snubber based soft-switching inverter. *Proc. Applied Power Electronics Conf.*, pp. 797–803.
18. Lai, J. S., Young Sr., R. W., Ott Jr., G.W., McKeever, J. W., and Peng, F. Z. 1996. A delta-configured auxiliary resonant snubber inverter. *IEEE Trans. Ind. Applicat.*, pp. 518–525.
19. Miller, T. J. E. 1989. *Brushless Permanent-Magnet and Reluctance Motor Drives*, Oxford, U.K.: Clarendon.
20. Sen, P. C. 1997. *Principles of Electric Machines and Power Electronics*. New York: John Wiley.
21. Divan, D. M., Venkataramanan, G., and De Doncker, R. W. 1987. Design methodologies for soft switched inverters. *Proc. IEEE Industry Applications Soc. Annu. Meeting*, pp. 626–639.
22. Yi, W., Liu, H. L., Jung, Y. C., Cho, J. G., and Cho, G. H. 1992. Program-controlled soft switching PRDCL inverter with new space vector PWM algorithm. *Proc. IEEE PESC*, pp. 313–319.
23. Ming, Z. Z. and Zhong, Y. R. 2001. A novel DC-rail parallel resonant ZVT VSI for three-phases ac motor drive. *Proc. Int. Conf. Electrontic Machines Sytems*, pp. 492–495.
24. Pan, Z. Y. and Luo, F. L. 2005. Transformer based resonant DC link inverter for brushless DC motor drive system. *IEEE Trans. Power Electron.*, pp. 939–947.
25. Sen, P. C. 1997. *Principles of Electric Machines and Power Electronics*. New York: John Wiley & Sons.
26. Wang, K. R., Jiang, Y. M., Dubovsky, S., Hua, G. C., Boroyevich, D., and Lee, F. C. 1997. Novel DC-rail soft-switched three-phase voltage-source inverters. *IEEE Tran. Ind. Applicat.*, pp. 509–517.

8

Multilevel DC/AC Inverters

Multilevel inverters use a different method to construct DC/AC inverters. This idea was published by Nabae in 1980 in an IEEE international conference, IEEE APEC'80 [1], and the same idea was published in 1981 in *IEEE Transactions on Industry Application* [2]. Actually, multilevel inverters employ a different technique from the PWM method, which vertically chops a reference waveform to achieve the similar output waveform (e.g., a sine wave). The multilevel inverting technique horizontally accumulates levels to achieve the waveform (e.g., a sine wave).

The soft-switching technique has been implemented in DC/DC conversion for more than 20 years. We introduce this technique in DC/AC inverters in this chapter.

8.1 Introduction

Although PWM inverters have been used in industrial applications, they have many drawbacks:

1. The carrier frequency must be very high. Mohan suggested $mf > 21$[3], which means $f\Delta > 1$ kHz if the output waveform has frequency 50 Hz. Usually, in order to keep the THD small, $f\Delta$ is selected as 2–20 kHz [3].

2. The pulse height is very high. In a normal PWM waveform (not multistage PWM), all pulse height is the DC linkage voltage. The output voltage of this PWM inverter has a large jumping span. For example, if the DC linkage voltage is 400 V, all pulses have the peak value 400 V. Usually, this causes large dv/dt and strong electromagnetic interference (EMI).

3. The pulse width would be very narrow when the output voltage has a low value. For example, if the DC linkage voltage is 400 V, the output is 10 V, and the corresponding pulse width should be 2.5% of the pulse period.

4. Terms 2 and 3 cause plenty of harmonics to produce poor THD.

5. Terms 2 and 3 result In a very rigorous switching condition. The switching devices experience large switching power losses.

6. The inverter control circuitry is complex, and the devices are costly. Therefore, the whole inverter is costly.

Multilevel inverters accumulate the output voltage in horizontal levels (layers). Therefore, using this technique overcomes the above drawbacks of the PWM technique:

1. The switching frequencies of most switching devices are low, which are equal to or only a small multiple of the output signal frequency.
2. The pulse heights are quite low. For an m-level inverter with output amplitude V_m, the pulse heights are V_m/m or only a small multiple of it. Usually, it causes low dv/dt and ignorable electromagnetic interference (EMI).
3. The pulse widths of all pulses have reasonable values that are comparable to the output signal.
4. Terms 2 and 3 cannot cause plenty of harmonics producing lower THD.
5. Terms 2 and 3 offer smooth switching condition. The switching devices have small switching power losses.
6. Inverter control circuitry is relatively simple, and the devices are not costly. Therefore, the inverter is economical.

Multilevel inverters contain several power switches and capacitors [4]. Output voltages of multilevel inverters are the sum of the voltages due to the commutation of the switches. Figure 8.1 shows one phase leg of inverters with different level numbers. A two-level inverter, as shown in Figure 8.1a, generates an output voltage with two levels with respect to the negative terminal of the capacitor. The three-level inverter shown in Figure 8.1b generates a three-level voltage, and an m-level inverter shown in Figure 8.1c generates an m-level voltage. Thus, the output voltages of multilevel inverters have several

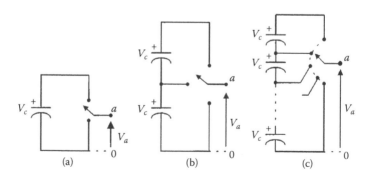

FIGURE 8.1
One phase leg of an inverter: (a) two levels, (b) three levels, and (c) m levels.

levels. Moreover, they can reach high voltage levels, while power semiconductors can withstand only lower voltages.

Multilevel inverters have been receiving increasing attention in recent decades, because of their many attractive features. Various kinds of multilevel inverters have been proposed, tested, and installed:

- Diode-clamped (neutral-clamped) multilevel inverters
- Capacitor-clamped (flying capacitors) multilevel inverters
- Cascaded multilevel inverters with separate DC sources
- H-bridge multilevel inverters
- Generalized multilevel inverters
- Mixed-level multilevel inverters
- Multilevel inverters by the connection of three-phase two-level inverters
- Soft-switched multilevel inverters
- Laddered inverters

The family tree of multilevel level inverters is shown in Figure 8.2.

The family of multilevel inverters has emerged the solution for high power applications [4]. It is hard to be implemented via a single power semiconductor switch directly in medium-voltage networks. Multilevel inverters have been applied to different high-power applications, such as large motor drives, railway traction applications, high-voltage DC transmissions (HVDC), unified power flow controllers (UPFC), static VAR compensators

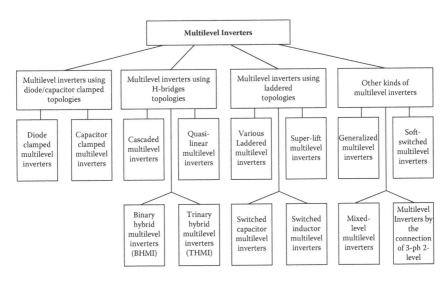

FIGURE 8.2
Family tree of multilevel inverters.

(SVCs), and static synchronous compensators (STATCOMs). The output voltage of the multilevel inverter has many levels synthesized from several DC voltage sources. The quality of the output voltage is improved as the number of voltage levels increases, so the effort of output filters can be decreased. The transformers can be eliminated due to reduced voltage that the life of the switch increases. Moreover, being cost-effective solutions, the applications of multilevel inverters are also extended to medium- and low-power applications such as electrical vehicle propulsion systems, active power filters (APFs), voltage sag compensations, photovoltaic systems, and distributed power systems.

Multilevel inverter circuits have been investigated for three decades. Separate DC-sourced full-bridge cells are connected in series to synthesize a staircase AC output voltage. The diode-clamped inverter, also called the neutral-point clamped (NPC) inverter, was presented in 1980 by Nabae. Because the NPC inverter effectively doubles the device voltage level without requiring precise voltage matching, this circuit topology prevailed in the 1980s. The capacitor-clamped (also called flying capacitor) multilevel inverter was introduced in the 1990s. Although the cascaded multilevel inverter was invented earlier, its application did not become widespread until the mid 1990s. The advantages of cascaded multilevel inverters have been indicated for motor drives and utility applications. The cascaded inverter has drawn great interest due to the great demand for medium-voltage high-power inverters.

The cascaded inverter is also used in regenerative-type motor drive applications. Recently, some new topologies of multilevel inverters have emerged, such as generalized multilevel inverters, mixed multilevel inverters, hybrid multilevel inverters, and soft-switched multilevel inverters. Today, multilevel inverters are extensively used in high-power applications with medium voltage levels, such as laminators, mills, conveyors, pumps, fans, blowers, compressors, and so on. Moreover, as a cost-effective solution, the applications of multilevel inverters are also extended to low-power applications such as photovoltaic systems, hybrid electrical vehicles, and voltage sag compensation, in which the effort of output filter components can be greatly decreased due to low harmonic distortions of output voltages of the multilevel inverters.

8.2 Diode-Clamped Multilevel Inverters

In this category, the switching devices are connected in series to make up the desired voltage rating and output levels. The inner voltage points are clamped by either two extra diodes or one high-frequency capacitor. The switching devices of an m-level inverter are required to block a voltage level of $V_{dc}/(m-1)$. The clamping diode needs different voltage ratings for different inner voltage levels. In summary, an m-level diode clamped inverter has

- Number of power electronic switches = $2(m - 1)$
- Number of DC-link capacitors = $(m - 1)$
- Number of clamped diodes = $2(m - 2)$
- Voltage across each DC-link capacitor = $\frac{V_{dc}}{m-1}$

where V_{dc} is the DC-link voltage. A three-level diode-clamped inverter is shown in Figure 8.3a with $V_{dc} = 2E$. In this circuit, the DC-bus voltage is split into three levels by two series-connected bulk capacitors, C_1 and C_2. The middle point of the two capacitors, n, can be defined as the neutral point. The output voltage v_{an} has three states: E, 0, and $-E$. For voltage level E, switches S_1 and S_2 need to be turned on; for $-E$, switches S_1' and S_2' need to be turned on; and for the 0 level, S_2 and $S_{2'}$ need to be turned on.

The key components that distinguish this circuit from a conventional two-level inverter are D_1 and $D_{1'}$. These two diodes clamp the switch voltage to

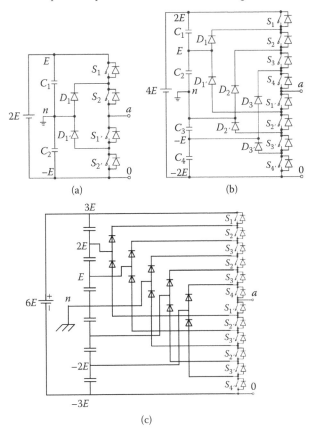

(a)

(b)

(c)

FIGURE 8.3
Diode-clamped multilevel inverter circuit topologies: (a) three-level, (b) five-level, and (c) seven-level.

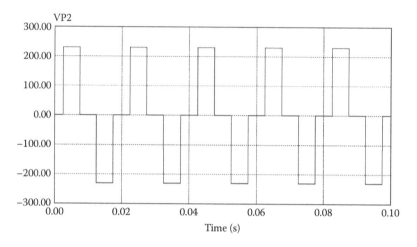

FIGURE 8.4
Output waveform of a three-level inverter.

half the level of the DC-bus voltage. When both S_1 and S_2 turn on, the voltage across a and 0 is $2E$, that is, $v_{a0} = 2E$. In this case, $D_{1'}$ balances out the voltage sharing between $S_{1'}$ and $S_{2'}$ with $S_{1'}$ blocking the voltage across C_1 and $S_{2'}$ blocking the voltage across C_2. Notice that output voltage v_{an} is AC, and v_{a0} is DC. The difference between v_{an} and v_{a0} is the voltage across C_2, which is E. If the output is removed between a and 0, then the circuit becomes a DC/DC converter that has three output voltage levels: E, 0, and $-E$. The simulation waveform is shown in Figure 8.4.

Usually, as the number of levels increases, the corresponding THD of the output voltage decreases. The switching angle decides the THD of the output voltage as well. The three-level diode clamped inverter has the THD is shown in Table 8.1.

Figure 8.3b) shows a five-level diode-clamped converter in which the DC bus consists of four capacitors: C_1, C_2, C_3, and C_4. For DC bus voltage $4E$, the voltage across each capacitor is E, and each device voltage stress will be limited to one capacitor voltage level E through clamping diodes.

TABLE 8.1

THD Content for Different Switching Angles ($m = 3$)

Methods	Switching Angle ($\alpha_1{}^\circ$)	THD
FFM	15°	31.76%
HHM	30°	30.9%

Note: FFM is a feed-forward method and HHM is a half-height method, which will be discussed in Chapter 14.

To explain how the staircase voltage is synthesized, the neutral point n is considered as the output phase voltage reference point. There are five switch combinations to synthesize five level voltage across a and n:

- For voltage level $v_{an} = 2E$, turn on all upper switches S_1–S_4.
- For voltage level $v_{an} = E$, turn on three upper switches S_2–S_4 and one lower switch $S_{1'}$.
- For voltage level $v_{an} = 0$, turn on two upper switches S_3 and S_4 and two lower switches $S_{1'}$ and $S_{2'}$.
- For voltage level $v_{an} = -E$, turn on one upper switch S_4 and three lower switches $S_{1'}$ –$S_{3'}$.
- For voltage level $v_{an} = -2E$, turn on all lower switches $S_{1'}$–$S_{4'}$.

For a diode-clamped inverter, each output level has only one combination to implement its output voltage. Four complementary switch pairs exist in each phase. The complementary switch pair is defined such that turning on one of the switches will exclude the other from being turned on. In this example, the four complementary pairs are $(S_1, S_{1'})$, $(S_2, S_{2'})$, $(S_3, S_{3'})$, and $(S_4, S_{4'})$. Although each active switching device is only required to block a voltage level of E, the clamping diodes must have different voltage ratings for reverse voltage blocking. Using $D_{1'}$ of Figure 8.3b as an example, when lower devices S_2–$S_{4'}$ are turned on, $D_{1'}$ needs to block three capacitor voltages, or $3E$. Similarly, D_2 and $D_{2'}$ need to block $2E$, and D_1 needs to block $3E$.

The simulation waveform is shown in Figure 8.5 below.

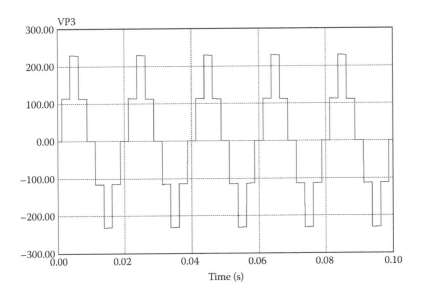

FIGURE 8.5
Output waveform of a five-level inverter.

TABLE 8.2

THD Content for Different Switching Angles ($m = 5$)

Methods	Switching Angle α_1 (°)	Switching Angle α_2 (°)	THD
FFM	7.24°	24.5°	24.86%
HHM	14.48°	49°	21.14%

The five-level diode-clamped inverter has the THD shown in Table 8.2.

A seven-level diode clamped inverter is shown in Figure 8.3c, and its output waveform is shown in Figure 8.6.

The seven-level diode-clamped inverter has the THD shown in Table 8.3.

From Figures 8.4 to 8.6, the THD is reduced when the number (m) of the level of the inverter is increased. Hence, a higher level inverter will produce output with less harmonic content. For each inverter, careful setting of the switching angles can obtain the best THD.

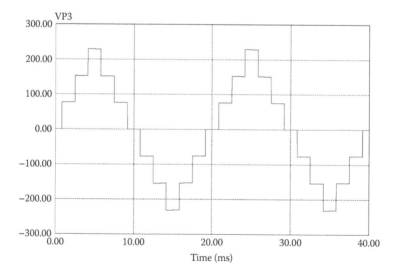

FIGURE 8.6

Output waveform of a seven-level inverter.

TABLE 8.3

THD Content for Different Switching Angles ($m = 7$)

Methods	α_1	α_2	α_3	THD
FFM	4.8°	15°	28.22°	22.17%
HHM	9.6°	30°	56.44°	11.70%

8.3 Capacitor-Clamped Multilevel Inverters (Flying Capacitor Inverters)

Figure 8.7 illustrates the fundamental building block of a phase-leg capacitor-clamped inverter. The circuit has been called the flying capacitor inverter with dependent capacitors clamping the device voltage to one capacitor voltage level. The inverter in Figure 8.7a provides a three-level output across a and n, that is, $v_{an} = E$, 0, or $-E$. For the voltage level E, switches S_1 and S_2 need to be turned on; for $-E$, switches $S_{1'}$ and $S_{2'}$ need to be turned on; and for the 0 level, either pair (S_1, $S_{1'}$) or (S_2, $S_{2'}$) needs to be turned on. Clamping capacitor C_1 is charged when S_1 and $S_{1'}$ are turned on, and is discharged when S_2 and $S_{2'}$ are turned on. The charge of C_1 can be balanced by proper selection of the 0-level switch combination.

The voltage synthesis in a five-level capacitor-clamped inverter has more flexibility than a diode-clamped converter. Using Figure 8.7b as an example, the voltage of the five-level phase-leg a output with respect to the neutral point n, v_{an}, can be synthesized by the following switching combinations:

- For voltage level $v_{an} = 2E$, turn on all upper switches S_1–S_4.
- For voltage level $v_{an} = E$, there are three combinations:
  - $S_1, S_2, S_3, S_{1'}$: $v_{an} = 2E$ (upper C_4) $- E$ (C_1);
  - $S_2, S_3, S_4, S_{4'}$: $v_{an} = 3E$ (C_3) $- 2E$ (lower C_4); and
  - $S_1, S_3, S_4, S_{3'}$: $v_{an} = 2E$ (upper C_4) $- 3E$ (C_3) $+ 2E$ (C_2).

- For voltage level $v_{an} = 0$, there are six combinations:
  - $S_1, S_2, S_{1'}, S_{4'}$: $v_{an} = 2E$ (upper C_4) $- 2E$ (C_2);
  - $S_3, S_4, S_{3'}, S_{4'}$: $v_{an} = 2E$ (C_2) $- 2E$ (lower C_4);
  - $S_1, S_3, S_{1'}, S_{3'}$: $v_{an} = 2E$ (upper C_4) $- 3E$ (C_3) $+ 2E$ (C_2) $- E$ (C_1);
  - $S_1, S_4, S_{2'}, S_{3'}$: $v_{an} = 2E$ (upper C_4) $- 3E$ (C_3) $+ E$ (C_1);
  - $S_2, S_4, S_{2'}, S_{4'}$: $v_{an} = 3E$ (C_3) $- 2E$ (C_2) $+ E$ (C_1) $- 2E$ (lower C_4); and
  - $S_2, S_3, S_{1'}, S_{4'}$: $v_{an} = 3E$ (C_3) $- E$ (C_1) $- 2E$ (lower C_4).

- For voltage level $V_{an} = -E$, there are three combinations:
  - $S_1, S_{1'}, S_{2'}, S_{3'}$: $v_{an} = 2E$ (upper C_4) $- 3E$ (C_3);
  - $S_4, S_{2'}, S_{3'}, S_{4'}$: $v_{an} = E$ (C_1) $- 2E$ (lower C_4); and
  - $S_3, S_{1'}, S_{3'}, S_{4'}$: $v_{an} = 2E$ (C_2) $- E$ (C_1) $- 2E$ (lower C_4).

- For voltage level $v_{an} = -2E$, turn on all lower switches, $S_{1'}$–$S_{4'}$.

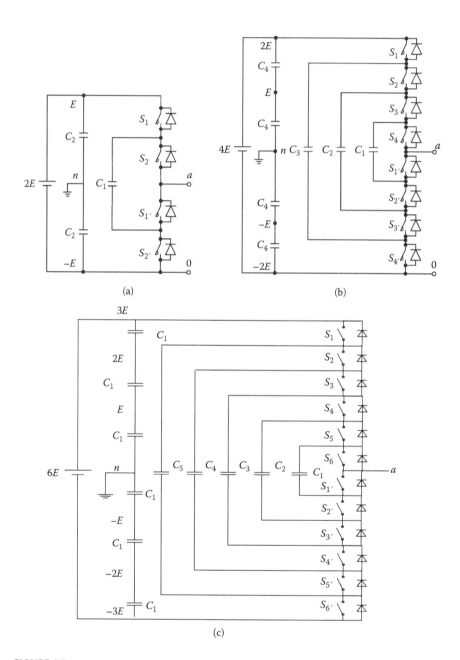

FIGURE 8.7
Capacitor-clamped multilevel inverter circuit topologies: (a) three-level, (b) five-level, and (c) seven-level.

Usually, the positive top level and negative top level have only one combination to implement their output values. Other levels have various combinations to implement their output values. In the preceding description, the capacitors with positive signs are in discharging mode, while those with negative sign are in charging mode. By proper selection of capacitor combinations, it is possible to balance the capacitor charge.

Figure 8.7c shows the seven-level capacitor-clamped inverter. The readers can synthesize the switching combinations for each output voltage level.

8.4 Multilevel Inverters Using H-Bridges (HBs) Converters

The basic structure is based on the connection of H-bridges (HBs). Figure 8.8 shows the power circuit for one phase leg of a multilevel inverter with three HBs (HB$_1$, HB$_2$, and HB$_3$) in each phase. Each HB is supplied by a separate DC source. The resulting phase voltage is synthesized by the addition of the voltages generated by the different HBs. If the DC link voltages of HBs are identical, the multilevel inverter is called the cascaded multilevel inverter. Its output waveform is shown in Figure 8.9. However, it is possible to have different values among the DC link voltages of HBs, and the circuit can be called a hybrid multilevel inverter.

Example 9.3. A three-HB multilevel inverter is shown in Figure 8.8. The output voltage is v_{an}, which is shown in Figure 8.10. It is implemented as

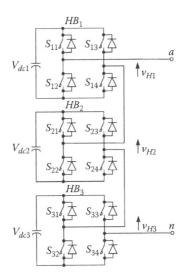

FIGURE 8.8
Multilevel inverter based on the connection of HBs.

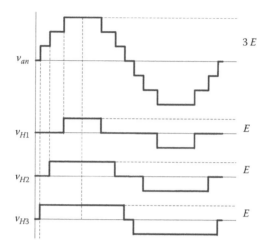

FIGURE 8.9
Waveforms of cascaded multilevel inverters.

a binary hybrid multilevel inverter (BHMI). Explain the inverter's working operation, and draw the corresponding waveforms, and indicate the source voltage's arrangement and how many levels can be implemented.

Solution: The DC link voltages of HB$_i$ (the ith HB), V_{dci}, is $2^{i-1}E$. In a 3-HB one phase leg,

$$V_{dc1} = E, \quad V_{dc2} = 2E, \quad V_{dc3} = 4E$$

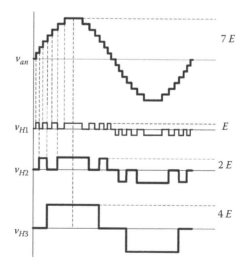

FIGURE 8.10
Waveforms of binary hybrid multilevel inverter (BHMI).

The operation is listed as follows:

+0: $v_{H1} = 0$, $v_{H2} = 0$, $v_{H3} = 0$,
+1E: $v_{H1} = E$, $v_{H2} = 0$, $v_{H3} = 0$,
+2E: $v_{H1} = 0$, $v_{H2} = 2E$, $v_{H3} = 0$,
+3E: $v_{H1} = E$, $v_{H2} = 2E$, $v_{H3} = 0$,
+4E: $v_{H1} = 0$, $v_{H2} = 0$, $v_{H3} = 4E$,
+5E: $v_{H1} = E$, $v_{H2} = 0$, $v_{H3} = 4E$,
+6E: $v_{H1} = 0$, $v_{H2} = 2E$, $v_{H3} = 4E$,
+7E: $v_{H1} = E$, $v_{H2} = 2E$, $v_{H3} = 4E$,
−E: $v_{H1} = -E$, $v_{H2} = 0$, $v_{H3} = 0$,
−2E: $v_{H1} = 0$, $v_{H2} = -2E$, $v_{H3} = 0$,
−3E: $v_{H1} = -E$, $v_{H2} = -2E$, $v_{H3} = 0$,
−4E: $v_{H1} = 0$, $v_{H2} = 0$, $v_{H3} = -4E$,
−5E: $v_{H1} = -E$, $v_{H2} = 0$, $v_{H3} = -4E$,
−6E: $v_{H1} = 0$, $v_{H2} = -2E$, $v_{H3} = -4E$,
−7E: $v_{H1} = -E$, $v_{H2} = -2E$, $v_{H3} = -4E$,

As shown in Figure 8.10, the output waveform, v_{an}, has 15 levels. One of the advantages is that the HB with higher DC link voltage has a lower number of commutations and thereby reduces the associated switching losses. The higher switching frequency components, for example, IGBT, are used to construct the HB with lower DC link voltage.

8.4.1 Cascaded Equal Voltage Multilevel Inverters (CEMI)

In the cascaded equal voltage multilevel inverter, the DC link voltages of HBs are identical, as shown in Figure 8.8.

$$V_{dc1} = V_{dc2} = V_{dc3} = E \tag{8.1}$$

where E is the unit voltage. Each HB generates three voltages at the output: $+E$, 0, and $-E$. This is made possible by connecting the capacitors sequentially to the AC side via the three power switches. The resulting output AC voltage swings from $-3E$ to $3E$ with seven levels as shown in Figure 8.9.

8.4.2 Binary Hybrid Multilevel Inverter (BHMI)

In the binary hybrid multilevel inverter (BHMI), the DC link voltages of HB_i (the ith HB), V_{dci}, is $2^{i-1}E$. In a 3-HB one phase leg,

$$V_{dc1} = E, \quad V_{dc2} = 2E, \quad V_{dc3} = 4E \tag{8.2}$$

As shown in Figure 8.10, the output waveform, v_{an}, has 15 levels. One of the advantages is that the HB with higher DC link voltage has fewer commutations, thereby reducing the associated switching losses. The BHMI illustrates a seven-level (in half-cycle) inverter using this hybrid topology. The HB with higher DC link voltage consists of the lower switching frequency component. The higher switching frequency components, for example, the IGBT, are used to construct the HB with lower DC link voltage.

8.4.3 Quasi-Linear Multilevel Inverter (QLMI)

In the quasi-linear multilevel inverter, the DC link voltages of HB_i, V_{dci} can be expressed as

$$
V_{dci} = \begin{cases} E & i = 1 \\ 2 \times 3^{i-2} E & i \geq 2 \end{cases} \tag{8.3}
$$

In a three-HB one-phase leg,

$$
V_{dc1} = E, \quad V_{dc2} = 2E, \quad V_{dc3} = 6E \tag{8.4}
$$

As shown in Figure 8.11, the output waveform, v_{an}, has 19 levels.

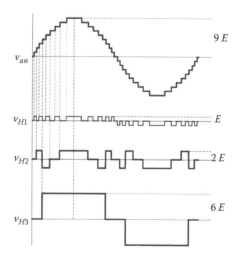

FIGURE 8.11
Waveforms of quasi-linear multilevel inverter.

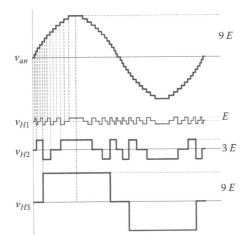

FIGURE 8.12
Waveforms of a 27-level trinary hybrid multilevel inverter (THMI).

8.4.4 Trinary Hybrid Multilevel Inverter (THMI)

In a trinary hybrid multilevel inverter, the DC link voltages of HB_i, V_{dci}, are $3^{i-1}E$. In a three-HB one phase leg,

$$V_{dc1} = E \quad V_{dc2} = 3E \quad V_{dc3} = 9E \tag{8.5}$$

As shown in Figure 8.12, the output waveform, v_{an}, has 27 levels. To the best of the authors' knowledge, this circuit has the greatest level for a given number of HBs among existing multilevel inverters.

8.5 Other Kinds of Multilevel Inverters

Several other kinds of multilevel inverters are introduced in this subsection [5–8].

8.5.1 Generalized Multilevel Inverters (GMI)

A generalized multilevel inverter topology has previously been presented. The existing multilevel inverters, such as diode-clamped and capacitor-clamped multilevel inverters, can be derived from this generalized multilevel inverter topology. Moreover, the generalized multilevel inverter topology can balance each voltage level by itself regardless of load characteristics. Therefore, the generalized multilevel inverter topology provides a true

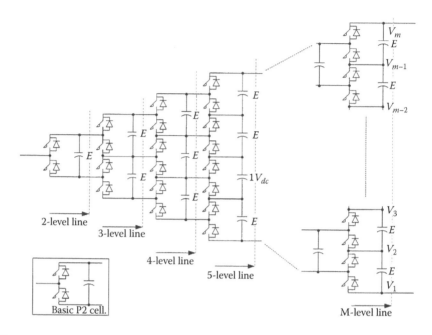

FIGURE 8.13
Generalized multilevel inverter structure.

multilevel structure that can balance each DC voltage level automatically at any number of levels, regardless of active or reactive power conversion, and without any assistance from other circuits. Thus, in principle, it provides a complete multilevel topology that embraces the existing multilevel inverters.

Figure 8.13 shows the generalized multilevel inverter structure per phase leg. Each switching device, diode, or capacitor's voltage is E, that is, $1/(m-1)$ of the DC link voltage. Any inverter with any number of levels, including the conventional two-level inverter, can be obtained using this generalized topology. As an application example, a four-level bidirectional DC/DC converter, shown in Figure 8.14, is suitable for the dual-voltage system to be adopted in future automobiles. The four-level DC/DC converter has a unique feature, which is that no magnetic components are needed. From this generalized multilevel inverter topology, several new multilevel inverter structures can be derived.

8.5.2 Mixed-Level Multilevel Inverter Topologies

For high-voltage high-power applications, it is possible to adopt multilevel diode-clamped or capacitor-clamped inverters to replace the full-bridge cell in a cascaded multilevel inverter. The reason for doing so is to reduce the number of separate DC sources. The nine-level cascaded inverter requires 4 separate DC sources for one phase leg and 12 for a three-phase inverter. If a three-level inverter replaces the full-bridge cell, the voltage level is effectively

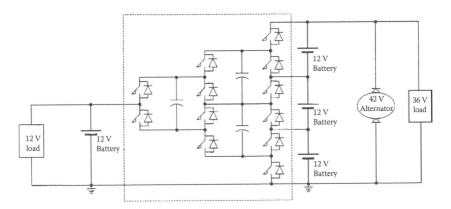

FIGURE 8.14
Application example: a four-level inverter for the dual-voltage system in automobiles.

doubled for each cell. Thus, to achieve the same nine voltage levels for each phase, only two separate DC sources are needed for one phase leg and six for a three-phase inverter. The configuration can be considered as having mixed-level multilevel cells because it embeds multilevel cells as the building block of the cascaded multilevel inverter.

8.5.3 Multilevel Inverters by Connection of Three-Phase Two-Level Inverters

Standard three-phase two-level inverters are connected by transformers as shown in Figure 8.15. In order for the inverter output voltages to be added

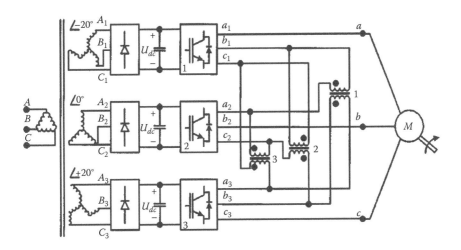

FIGURE 8.15
Cascaded inverter with three-phase cells.

up, the inverter outputs of the three modules need to be synchronized with a separation of $120°$ between each phase. For example, obtaining a three-level voltage between outputs a and b, the voltage is synthesized by $V_{ab} = V_{a1-b1} + V_{a1-b1} + V_{a1-b1}$. The phase between b_1 and a_2 is provided by a_3 and b_3 through an isolated transformer. With three inverters synchronized, the voltages V_{a1-b1}, V_{a1-b1}, and V_{a1-b1} are all in phase; thus, the output level is simply tripled.

References

1. Nabae, A., Takahashi, I., and Akagi, H. 1980. A neutral-point clamped PWM inverter. *Proc. IEEE APEC'80 Conf.*, pp. 761–766.
2. Nabae, A., Takahashi, I., and Akagi, H. 1981. A neutral-point clamped PWM inverter. *IEEE Trans. Ind. Applicat.*, pp. 518–523.
3. Mohan, N., Undeland, T. M., and Robbins, W. P. 2003. *Power Electronics: Converters, Applications and Design*. New York: John Wiley & Sons.
4. Liu, Y. and Luo, F. L. 2008. Trinary hybrid 81-level multilevel inverter for motor drive with zero common-mode voltage. *IEEE Trans. Ind. Electron.*, pp. 1014–1021.
5. Manjrekar, M. D., Steimer, P. K., and Lipo, T. A. 2000. Hybrid multilevel power conversion system: a competitive solution for high-power applications. *IEEE Trans. Ind. Applicat.*, pp. 834–841.
6. Akagi, H. 2006. Medium-voltage power conversion systems in the next generation. *Proc. IEEE-IPEMC'2006*, pp. 23–30.
7. Inoue, S. and Akagi, H. 2007. A bidirectional isolated DC–DC converter as a core circuit of the next-generation medium-voltage power conversion system. *IEEE Trans. Power Electron.*, pp. 535–542.
8. Liu, Y. and Luo, F. L. 2006. Multilevel inverter with the ability of self voltage balancing. *IEE Proc. Electric Power Applicat.*, pp. 105–115.

9

Trinary Hybrid Multilevel Inverter (THMI)

The trinary hybrid multilevel inverter (THMI) has many advantages. We carefully analyze its characteristics in this chapter [1].

9.1 Topology and Operation

A single-phase THMI with h HBs connected in series is shown in Figure 9.1. The key feature of the THMI is that the ratio of DC link voltage is $1:3:\ldots:3^{h-1}$, where h is the number of HBs. The maximum number of synthesized voltage levels is 3^h.

As shown in Figure 9.1, v_{Hi} represents the output voltage of the ith HB. V_{dci} represents the DC link voltage of the ith HB. A switching function, F_i, is used to relate V_{Hi} and V_{dci} as shown in

$$v_{Hi} = F_i \cdot V_{dci} \tag{9.1}$$

The value of F_i can be either 1 or -1 or 0. For the value 1, switches S_{i1} and S_{i4} need to be turned on. For the value -1, switches S_{i2} and S_{i3} need to be turned on. For the value 0, switches S_{i1} and S_{i3} need to be turned on or S_{i2} and S_{i4} need to be turned on. Table 9.1 shows the relationship of the switching function, the output voltage of an HB, and states of switches.

The output voltage of the THMI, v_{an}, is the sum of the output voltages of HBs.

$$v_{an} = \sum_{i=1}^{h} v_{Hi} \tag{9.2}$$

From Equations (9.1) and (9.2), we can get

$$v_{an} = \sum_{i=1}^{h} F_i \cdot V_{dci} \tag{9.3}$$

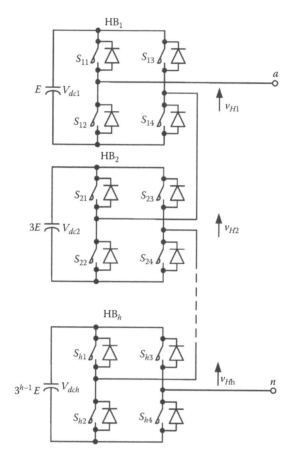

FIGURE 9.1
Configuration of THMI.

TABLE 9.1

Relationship of Switching Function, Output Voltage
of an HB, and States of Switches

F_i	v_{Hi}	S_{i1}	S_{i2}	S_{i3}	S_{i4}
1	V_{dci}	Conduct	Block	Block	Conduct
−1	$-V_{dci}$	Block	Conduct	Conduct	Block
0	0	Conduct	Conduct	Block	Block
0	0	Block	Block	Conduct	Conduct

In a single-phase h-HB THMI, the ratio of DC link voltages is $1{:}3{:}\ldots{:}3^{h-1}$. Suppose E is the unit voltage, the DC link voltage can be expressed as

$$V_{dci} = 3^{i-1}E \tag{9.4}$$

From Equations (9.2) and (9.3), we can get

$$v_{an} = \sum_{i=1}^{h} F_i \cdot 3^{i-1}E \tag{9.5}$$

Suppose l is the ordinal of the expected voltage level that the inverter outputs. If l is not negative, the inverter outputs the positive lth voltage level. If l is negative, the inverter outputs the negative ($-l$)th voltage level. In a single-phase THMI with h HBs, given the value of l, the value of F_i can be determined by

$$F_h = \frac{\mathrm{ABS}(l)}{l} B_b \left(\mathrm{ABS}(l) - \frac{3^{h-1} - 1}{2} \right)$$

$$F_{h-1} = \frac{\mathrm{ABS}(l)}{l} B_b \left(\mathrm{ABS}(l) - \mathrm{ABS}(F_h) \cdot 3^{h-1} - \frac{3^{h-2} - 1}{2} \right)$$

$$\vdots$$

$$F_i = \frac{\mathrm{ABS}(l)}{l} B_b \left(\mathrm{ABS}(l) - \sum_{k=i+1}^{h} \left(\mathrm{ABS}(F_k) \cdot 3^{k-1} \right) - \frac{3^{i-1} - 1}{2} \right) \tag{9.6}$$

$$\vdots$$

$$F_2 = \frac{\mathrm{ABS}(l)}{l} B_b \left(\mathrm{ABS}(l) - \sum_{k=3}^{h} \left(\mathrm{ABS}(F_k) \cdot 3^{k-1} \right) - 1 \right)$$

$$F_1 = \frac{\mathrm{ABS}(l)}{l} B_b \left(\mathrm{ABS}(l) - \sum_{k=2}^{h} \left(\mathrm{ABS}(F_k) \cdot 3^{k-1} \right) \right)$$

where ABS is the function of absolute value and the bipolar binary function, B_b, is defined as

$$B_b(\tau) = \begin{cases} 1 & \tau > 0 \\ 0 & \tau = 0 \\ -1 & \tau < 0 \end{cases} \tag{9.7}$$

TABLE 9.2

Relationship of Output Voltage of
the Inverter and the Values of
Switching Functions in a
Single-Phase Two-HB THMI

v_{an}	$-4E$	$-3E$	$-2E$	$-E$	0
F_1	-1	0	1	-1	0
F_2	-1	-1	-1	0	0
v_{an}	$4E$	$3E$	$2E$	E	
F_1	1	0	-1	1	
F_2	1	1	1	0	

From Equation (9.5), we can get the relationship between the output voltage of the inverter, v_{an}, and the values of switching functions in the THMI with different numbers of HBs. In the case of a two-HB THMI, Table 9.2 shows the relationship between the output voltage of the inverter and the values of switching functions. The waveforms of a single-phase two-HB THMI are shown in Figure 9.2.

The output voltage of a single-phase three-HB has 27 levels. v_{H1}, v_{H2}, and v_{H3} can be negative when v_{an} is positive. Table 9.3 shows relationship between the output voltage of the inverter and the values of switching functions in a single-phase three-HB THMI. From Equation (9.6), we can get

$$v_{an} = -v'_{an} \Leftrightarrow F_i = F'_i \quad i = 1 \dots h \tag{9.8}$$

The cases with negative values of v_{an} can be deduced from Table 9.3.

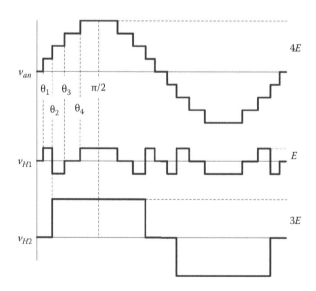

FIGURE 9.2
Waveforms of a single-phase two-HB THMI.

TABLE 9.3

Relationship of Output Voltage of the Inverter
and the Values of Switching Functions in a
Single-phase three-HB THMI

v_{an}	13E	12E	11E	10E	9E	8E	7E
F_1	1	0	−1	1	0	−1	1
F_2	1	1	1	0	0	0	−1
F_3	1	1	1	1	1	1	1
v_{an}	6E	5E	4E	3E	2E	E	0
F_1	0	−1	1	0	−1	1	0
F_2	−1	−1	1	1	1	0	0
F_3	1	1	0	0	0	0	0

9.2 Proof of Greatest Number of Output Voltage Levels

Among existing multilevel levels, the THMI has the greatest levels of output voltage using the same number of components. In this section, first this assertion is theoretically proved, and then various kinds of multilevel inverters are compared.

9.2.1 Theoretical Proof

This section proves that the THMI has greatest levels of output voltage using the same number of HBs among the multilevel inverters using HBs connected. A phase voltage waveform is obtained by summing the output voltages of h HBs as shown in Equation (9.2). If the DC link sources of all HB cells are equal, the multilevel inverter is called the cascaded multilevel inverter and the maximum number of levels of phase voltage is given by

$$m = 1 + 2h \qquad (9.9)$$

On the other hand, if at least one of the DC link sources is different from the other ones, the multilevel inverter is called the hybrid multilevel inverter. Thus, considering that the lowest DC link source E is chosen as the base value for the p.u. notation, the normalized values of all DC link voltages must be natural numbers to obtain a uniform step multilevel inverter, that is,

$$V_{dci^*} \in E, \qquad i = 1, 2, \dots, h \qquad (9.10)$$

Moreover, to obtain a uniform step multilevel inverter, the DC link voltage of the HB cells must also respect the following relation:

$$V_{dci^*} \leq 1 + 2 \sum_{k=1}^{i-1} V_{dck^*}, \qquad i = 2, 3, \dots, h \qquad (9.11)$$

where it is also considered that the DC link voltages are arranged in ascending order, that is,

$$V_{dc1*} \leq V_{dc2*} \leq V_{dc3*} \leq \cdots \leq V_{dch*} \tag{9.12}$$

Therefore, the maximum number of levels of output phase voltage waveform can be given as

$$m = 1 + 2\sigma_{max} \tag{9.13}$$

where σ_{max} is the maximum number of positive and negative voltage levels and can be expressed as

$$\sigma_{max} = \sum_{i=1}^{h} V_{dci*} \tag{9.14}$$

From Equations (9.9), (9.13), and (9.14), it is possible to verify that hybrid multilevel inverters can generate a large number of levels with the same number of cells. Moreover, in the THMI, the DC link voltages obey the relation

$$V_{dci*} = 1 + 2\sum_{k=1}^{i-1} V_{dck*}, \quad i = 2,3,\dots,h \tag{9.15}$$

Therefore, the THMI has the greatest levels of output voltages using the same number of HBs among multilevel inverters with HBs connected.

9.2.2 Comparison of Various Kinds of Multilevel Inverters

Two kinds of comparisons are presented in this section. In the first comparison, the components are considered to have same voltage rating, E. This comparison is for high-power and high-voltage applications, in which devices connected in series are used to satisfy the requirement of high voltage ratings. Table 9.4 shows the comparison of multilevel inverters: diode-clamped multilevel inverter (DCMI), capacitor-clamped multilevel inverter (CCMI), cascaded multilevel inverter (CMI), generalized multilevel inverter (GMI), BHMI,

TABLE 9.4

First Comparison of Multilevel Inverters

Converter Type	DCMI	CCMI	GMI	CMI	BHMI	THMI
Main switching devices	$2m-2$	$2m-2$	2^m-2	$2m-2$	$2m-2$	$2m-2$
Diodes	$m(m-1)$	$m-1$	2^m-2	$2m-2$	$2m-2$	$2m-2$
Capacitors	$m-1$	$0.5m(m-1)$	$m-1$	$(m-1)/2$	$(m-1)/2$	$(m-1)/2$
Total components	$(m-1)(m+1)$	$(m-1)(0.5m+3)$	$2^{m+1}+m-5$	$4.5(m-1)$	$4.5(m-1)$	$4.5(m-1)$

TABLE 9.5

Second Comparison of Multilevel Inverters

Converter Type	DCMI	CCMI	CMI	GMI	BHMI	THMI
Main switching devices	$2m-2$	$2m-2$	$2m-2$	2^m-2	$4 \times \log_2[(m+1)/2]$	$4 \times \log_3 m$
Diodes	$4m-6$	$2m-2$	$2m-2$	2^m-2	$4 \times \log_2[(m+1)/2]$	$4 \times \log_3 m$
Capacitors	$m-1$	$2m-3$	$0.5m-0.5$	$m-1$	$\log_2[(m+1)/2]$	$\log_3 m$
Total components	$7m-9$	$6m-7$	$4.5m-4.5$	$2^{m+1}+m-5$	$9 \times \log_2[(m+1)/2]$	$9 \times \log_3 m$

and THMI. m is the number of steps of phase voltage. From Table 9.4, we can see that CMI, BHMI, and THMI use fewer components. CMI, BHMI, and THMI use the same number of components. However, in practical systems, the redundancy requirement must be satisfied. THMI uses fewer components than BHMI and CMI in practical systems since THMI uses fewer redundant components. Moreover, THMI uses fewer DC sources than CMI and BHMI.

The second comparison is for medium- and low-power applications, in which the voltage rating of main switching components, diodes, and capacitors can be researched easily. Therefore, the numbers of main switching components, diodes, and capacitors are the minimal required values. Table 9.5 shows the comparison results among DCMI, CCMI, CMI, GMI, BHMI, and THMI. From Table 9.5, we can see that THMI uses the fewest components among these multilevel inverters.

9.2.3 Modulation Strategies for THMI

Five modulation strategies for THMI are investigated. They are the step modulation strategy, the virtual stage modulation strategy, the hybrid modulation strategy, the sub-harmonics pulse width modulation (PWM) strategy, and the simple modulation strategy. Since multilevel inverters are typically used in three-phase systems, generally only modulation strategies for three-phase systems will be investigated here. In three-phase systems, triple-order harmonic components of voltages need not be eliminated by the modulation strategies since they can be eliminated by proper connection of three-phase voltage sources and loads. In other words, only the 5th, 7th, 11th, 13th, 17th, 19th, ..., harmonic components should be eliminated by the modulation strategies. In addition, the amplitude of the fundamental component should be controlled. The list can be expressed by

$$\eta_i = \begin{cases} 3i-2 & \forall i = \text{odd} \\ 3i-1 & \forall i = \text{even} \end{cases} \quad i > 0 \tag{9.16}$$

The step modulation strategy, the virtual stage modulation strategy, and the simple modulation strategy are low-frequency modulation strategies. The

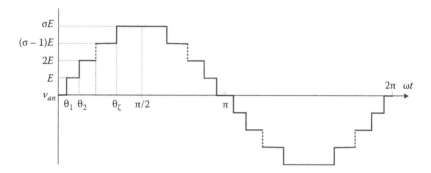

FIGURE 9.3
Step modulation strategy of THMI.

high-frequency modulation strategies used in hybrid multilevel inverters include the hybrid modulation strategy and the subharmonic PWM strategy.

9.2.3.1 Step Modulation Strategy

Figure 9.3 shows a general quarter-wave symmetric stepped voltage waveform synthesized by a THMI where E indicates the unit voltage of DC source. Consider that ς is the number of switching angles in a quarter wave of v_{an} and σ is the number of positive and negative levels of v_{an}. In the step modulation strategy,

$$\varsigma = \sigma \tag{9.17}$$

By applying Fourier series analysis, the amplitude of any odd jth harmonic of v_{an} can be expressed as

$$|v_{an}|_j = \frac{4}{j\pi} \sum_{i=1}^{\varsigma} [E\cos(j\theta_i)] \tag{9.18}$$

where j is an odd harmonic order and θ_i is the ith switching angle. The amplitudes of all even harmonics are zero. According to Figure 9.3, θ_1 to θ_ς must satisfy

$$0 < \theta_1 < \theta_2 < ... < \theta_\varsigma < \pi/2 \tag{9.19}$$

$$\begin{cases} \displaystyle\sum_{i=1}^{\varsigma} \cos(\eta_1\theta_i) = \sigma \cdot MR \\[2mm] \displaystyle\sum_{i=1}^{\varsigma} \cos(\eta_2\theta_i) = 0 \\[2mm] \quad\quad \vdots \\[2mm] \displaystyle\sum_{i=1}^{\varsigma} \cos(\eta_\varsigma\theta_i) = 0 \end{cases} \tag{9.20}$$

where *MR* is the relative modulation index and is expressed as

$$MR = \frac{\pi |v_{an}|_1}{4\sigma E} \qquad (9.21)$$

where $|v_{an}|_1$ is the amplitude of fundamental component of the output voltage of the inverter.

The switching angles controlled by step modulation technique are derived from Equation (9.20). Up to $(\varsigma - 1)$ harmonic contents can be removed from the voltage waveform and the amplitude of fundamental component can be controlled.

$$\cos(\theta_1) + \cos(\theta_2) + \cos(\theta_3) + \cos(\theta_4) = 0.83 \times 4$$

$$\cos(5\theta_1) + \cos(5\theta_2) + \cos(5\theta_3) + \cos(5\theta_4) = 0$$

$$\cos(7\theta_1) + \cos(7\theta_2) + \cos(7\theta_3) + \cos(7\theta_4) = 0 \qquad (9.22)$$

$$\cos(11\theta_1) + \cos(11\theta_2) + \cos(11\theta_3) + \cos(11\theta_4) = 0$$

The equation sets (Equation 9.20) from which the switching angles can be derived are nonlinear and transcendental. For example, in a two-HB THMI, with the step modulation technique, the equations set is expressed as Equation (9.22) when the relative modulation index is 0.83. The correct solution must satisfy the inequality shown in Equation (9.19).

The constrained optimization approach can be used to solve the nonlinear and transcendental equations sets. Each equation is regarded as an equational constraint. However, the computational problems of constrained optimization do not converge easily. Since in the actual electric system there are always mismatches and parameter tolerances, lower-order harmonics will be small but not exactly zero. This gives rise to the idea of transforming the constraint optimization model to a nonconstraint one. The nonconstraint optimization is expected to have a better convergence property.

The target function of the new scheme of optimization without equational constraints can be written as

$$FT = p_1 \left[\sum_{i=1}^{\varsigma} \cos(\eta_1 \theta_i) - \sigma \cdot M \right]^2 + p_2 \left[\sum_{i=1}^{\varsigma} \cos(\eta_2 \theta_i) \right]^2 + \cdots + p_\varsigma \left[\sum_{i=1}^{\varsigma} \cos(\eta_\varsigma \theta_i) \right]^2 \qquad (9.23)$$

$p_1 \sim p_\varsigma$ are penalty factors. The penalty factors were selected as

$$p_i = \frac{4}{2i-1} \qquad (9.24)$$

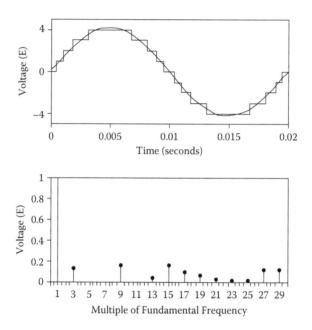

FIGURE 9.4
Synthesized phase leg voltage waveform and frequency spectrum of a two-HB THMI with step modulation technique.

Thus, the penalty factors put more weight on elimination of lower-order harmonics. Function f_{mincon} in the MATLAB® optimization toolbox was used to solve this minimization problem.

The two-HB THMI can synthesize a nine-level output voltage. Figures 9.4 and 9.5 show the typical synthesized waveform of the phase leg voltage, line-to-line voltage waveform and their frequency spectrums, as MR is equal to 0.83. The switching angles are 0.1478, 0.3232, 0.5738, and 0.9970. According to Equation (9.20), the fifth, seventh, and eleventh harmonics of the phase leg voltage can be eliminated in the two-HB THMI as shown in Figure 9.4. The THD of phase leg voltage is 9.66%. The triple-order harmonic components do not exist in the line-to-line voltage as shown in Figure 9.5. The THD of line-to-line voltage is 5.91%.

According to Equation (9.20), all switching angles must satisfy the constraint (Equation 9.19). If switching angles do not satisfy the constraint, this scheme no longer exists. The theoretical maximum amplitude of the fundamental component is $4\varsigma E/\pi$, which occurs as θ_1–θ_h equal zero. Because of the internal restriction of switching angles, the relative modulation index has upper and lower limitations. The limitation of the relative modulation index can be explained using Figures 9.6 and 9.7.

As shown in Figure 9.6, as the relative modulation index is less than a certain value, denoted by MR (min), θ_ς approaches $\pi/2$ and the limitation of

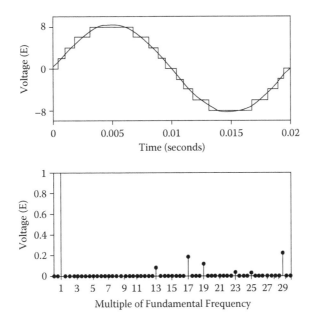

FIGURE 9.5
Synthesized line-to-line voltage waveform and frequency spectrum of a two-HB THMI with step modulation technique.

minimum modulation index occurs. Similarly, when the relative modulation index is greater than MR (max), θ_1 approaches zero and the limitation of the maximum modulation index occurs as shown in Figure 9.7.

For a THMI with h HBs, the maximum number of levels of the phase leg voltage is m, which equals 3^h. The maximum number of the positive/negative phase leg voltage levels is σ_{max}, which equals $(m{-}1)/2$. As mentioned above, the relative modulation index MR has limitations. To extend it to the smaller

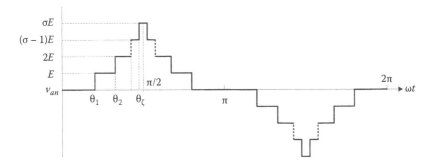

FIGURE 9.6
Limitation to the minimum MR in the step modulation.

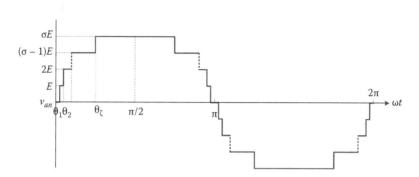

FIGURE 9.7
Limitation to the maximum *MR* in the step modulation

ranges of the modulation index, the inverter will output fewer voltage levels. Consequently, the number of positive/negative voltage levels that the inverter outputs, σ, is smaller than the maximum number of the positive/negative levels, σ_{max}. In the step modulation strategy, the number of switching angles in the quarter wave of v_{an}, ς, equals σ. The definition of the relative modulation index, *MR*, is based on σ as shown in Equation (9.21). This definition is easily included in Equation (9.20) to express the nonlinear transcendental equation sets that are used to calculate the switching angles. In practice, the modulation index, *M*, is used. *M* is based on the σ_{max} and can be expressed as

$$M = \frac{\pi \, |v_{an}|_1}{4\sigma_{max}E} \tag{9.25}$$

The relationship between *MR* and *M* can be expressed as

$$\frac{M}{MR} = \frac{\sigma}{\sigma_{max}} \tag{9.26}$$

In the two-HB THMI, according to Equation (9.20), the maximum *MR* is calculated as 0.86 and the minimum *MR* is 0.55 as the levels of output voltage are nine in number. The range of *M* is also from 0.55 to 0.86 with the nine-level output voltage. To extend the lower modulation index, fewer output voltage levels are synthesized. The range of *MR* is 0.46–0.83 when the number of output voltage levels is seven. According to Equation (9.26), the range of *M* is 0.34–0.62 when there are seven output voltage levels. Thus, the modulation range is extended to 0.34 by decreasing levels of output voltage.

Table 9.6 shows the relative modulation index and the modulation index with different output voltage levels in the two-HB THMI. First, the minimum and maximum *MR* is calculated by the optimization method. Second, the minimum and maximum *M* is calculated by Equation (9.26). It is preferable to use more output voltage levels. The last column of Table 9.6 shows the arrangement of *M* with different output voltage levels. In addition, the upper

TABLE 9.6

Range of Modulation Index under Different Output Voltage
Levels with Step Modulation in a Two-HB THMI

σ	MR (min)	MR (max)	M (min)	M (max)	Range of M
1	0	1	0	0.25	0-0.15
2	0.3	0.9	0.15	0.45	0.15–0.34
3	0.46	0.83	0.34	0.63	0.34–0.55
4	0.55	0.86	0.55	0.86	0.55–0.86
4*	0.3	0.94	0.3	0.94	0.86–0.94

* Upper limit of M without elimination of 11th harmonie.

limit of M can reach 0.94 independent of the elimination of the 11th harmonic
as shown in the last row of Table 9.6.

The scheme of switching angles of the two-HB THMI is shown in Figure 9.8.
When the modulation index reaches the lower limit, such as 0.34, the third
switching angle is close to $\pi/2$, which corroborates Figure 9.6. When the modulation index reaches the maximum value 0.86 or 0.94, the first angle is close
to zero, which corroborates Figure 9.7.

9.2.3.2 Virtual Stage Modulation Strategy

In the step modulation strategy, the output voltage levels of the multilevel
inverter limit the number of eliminated lower-order harmonics. Only three

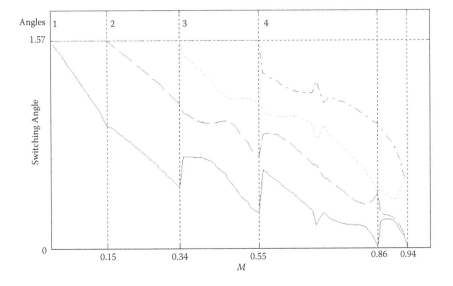

FIGURE 9.8
Scheme of switching angles with the step modulation as a function of modulation index in a
two-HB THMI.

lower-order harmonics can be eliminated by the step modulation in a two-HB THMI. It is not generally satisfied in the applications that require a high-quality sinusoid voltage output. The virtual stage modulation strategy is a new technique that increases the amount of eliminated lower-order harmonics without increasing the number of output voltage levels. The switching angles can be derived as

$$
\begin{cases}
\displaystyle\sum_{i=1}^{\alpha}\cos(\eta_1\theta_{pi}) - \sum_{i=1}^{\beta}\cos(\eta_1\theta_{ni}) = \sigma \cdot MR \\[2mm]
\displaystyle\sum_{i=1}^{\alpha}\cos(\eta_2\theta_{pi}) - \sum_{i=1}^{\beta}\cos(\eta_2\theta_{ni}) = 0 \\[1mm]
\vdots \\[1mm]
\displaystyle\sum_{i=1}^{\alpha}\cos(\eta_\varsigma\theta_{pi}) - \sum_{i=1}^{\beta}\cos(\eta_\varsigma\theta_{ni}) = 0
\end{cases}
\tag{9.27}
$$

where σ is the number of positive/negative levels of v_{an} and can be expressed as

$$\sigma = \alpha - \beta \tag{9.28}$$

where ς is the number of switching angles in a quarter waveform of v_{an} and can be expressed as

$$\varsigma = \alpha + \beta \tag{9.29}$$

MR is shown in Equation (9.21). Equation (9.27) is subject to

$$
\begin{cases}
0 < \theta_{p1} < \theta_{p2} < \ldots < \theta_{p\alpha} < \pi/2 \\
0 < \theta_{n1} < \theta_{n2} < \ldots < \theta_{n\beta} < \pi/2 \\
\theta_{nj} < \theta_{p(j+\sigma)} \qquad j = 1,2,\ldots,\beta
\end{cases}
\tag{9.30}
$$

In the two-HB THMI, when the output voltage changes between E and $2E$ or $-E$ and $-2E$, the switching components of the higher voltage HB will switch on and off as shown in Figure 9.2. To keep high-voltage switching components at a lower frequency in the THMI, the limitation in Equation (9.31) is added into Equation (9.27) to ensure that higher-voltage switching components switch at the fundamental frequency.

$$\theta_{p2} < \theta_{n1} \tag{9.31}$$

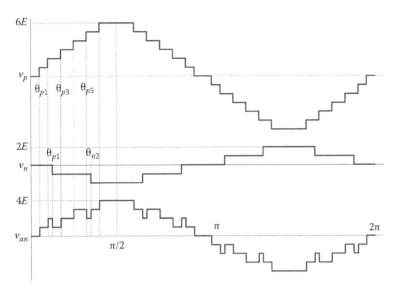

FIGURE 9.9
Waveform using the virtual stage modulation two-HB, nine-level, $\alpha = 6$, $\beta = 2$.

Figure 9.9 illustrates the waveform using the virtual stage modulation for the two-HB THMI with nine output voltage levels. The number of virtual stages, β, is two.

Figures 9.10 and 9.11 show the typical synthesized waveform of the phase leg voltage, line-to-line voltage waveform and their frequency spectrum in the virtual stage modulation strategy. The MR is 0.83, and the number of virtual stages is two. θ_{p1} to θ_{p6} are 0.1321, 0.3320, 0.5307, 0.6226, 0.9133, and 1.0419. θ_{n1} is 0.5750, and θ_{n2} is 0.9652. Because of two additional virtual stages, four more degrees of freedom of switching angles are created such that the 13th, 17th, 19th, and 23rd harmonics can be eliminated from the phase leg voltage as shown in Figure 9.10. The THD of the phase leg voltage is 10.67%. The triple-order harmonic components of line-to-line voltage do not exist, and the harmonics are pushed to 1250 Hz as shown in Figure 9.11. The THD of the line-to-line voltage is 7.3%.

In virtual stage modulation strategy, the relative modulation index also has upper and lower limits. Compared with the step modulation strategy, the optimal computation of the virtual stage modulation strategy tolerates more unequal restriction, as shown in Equations (9.30) and (9.31). When the switching angles do not satisfy these restrictions, the themes of switching angles no longer exist.

The concept of the relative modulation index can be used in the step modulation strategy by a similar method. Table 9.7 shows two cases. One is the nine-level output voltage with two virtual stages, and the other is the seven-level output voltage with one virtual stage. With the nine-level output voltage and two virtual stages, the 5th, 7th, 11th, 13th, 17th, 19th, and 23rd harmonics

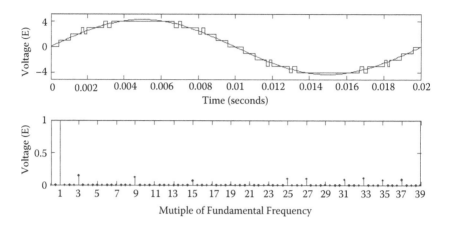

FIGURE 9.10
Synthesized phase leg voltage waveform and frequency spectrum of a two-HB THMI with virtual stage modulation.

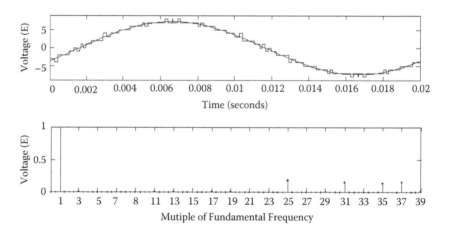

FIGURE 9.11
Synthesized line-to-line voltage waveform and frequency spectrum of a two-HB THMI with virtual stage modulation.

TABLE 9.7

Range of Modulation Indexes with Virtual Stage Modulation in a Two-HB THMI

σ	β	MR (min)	MR (max)	M (min)	M (max)	Range of M
3	1	0.51	0.92	0.38	0.69	0.38–0.459
4	2	0.459	0.92	0.459	0.92	0.459–0.92

Note: Indices in a two-HB THMI: p1 to p6 = θ_{p1} to θ_{p6}, n1 to n2 = θ_{n1} to θ_{n2}.

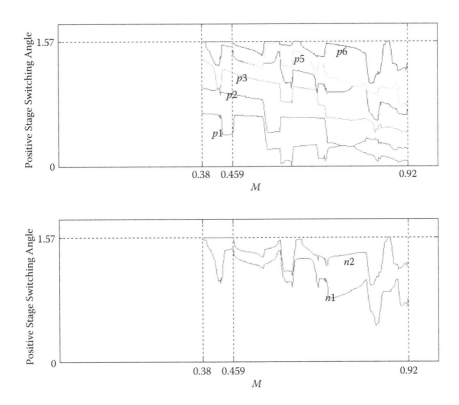

FIGURE 9.12
Scheme of switching angles for virtual stage modulation as a function of modulation.

can be eliminated. With the seven-level output voltage and one virtual stage, the 5th, 7th, 11th, and 13th harmonics can be eliminated. When there are five or three output voltage levels, the virtual stage modulation strategy is not applicable in the two-HB THMI since the restriction in Equation (9.31) must be violated. Therefore, when M is less than 0.38 in this case, the step modulation strategy will be used. With the virtual stage modulation strategy, the scheme of switching angles is shown in Figure 9.12.

9.2.3.3 Hybrid Modulation Strategy

The hybrid modulation strategy for the hybrid multilevel inverters has been presented, which incorporates stepped voltage waveform synthesis in higher-power HB cells in conjunction with high-frequency variable PWM in the lowest-power HB cell. Figure 9.13 presents a block diagram of the command circuit utilized to determine the command signals of the power devices of all HBs. As shown in Figure 9.13, the reference signal of the hybrid multilevel inverter, v_{ref} is the command signal of the HB with the highest DC voltage source ($V_{dc,h}$). This signal is compared with a voltage level corresponding to the

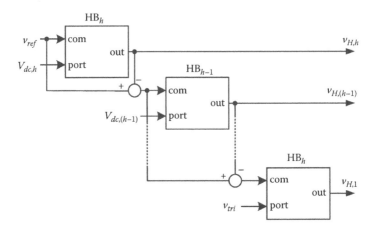

FIGURE 9.13
Hybrid modulation for hybrid multilevel inverters.

sum of all smaller DC voltage sources of the hybrid multilevel inverter, $\sigma_{max,h-1}$. If the command signal is greater than this level, the output of the inverter with the highest DC voltage source must be equal to $V_{dc,h}$. In addition, if the command signal is less than the negative value of $\sigma_{max,h-1}$, the output of this cell must be equal to $-V_{dc,h}$, otherwise the output of this cell must be zero.

The command signal of the ith HB cell is the difference between the command signal of the HB_{i+1} and the output voltage of the HB_{i+1}. In this way, the command signal of the ith cell contains information about the harmonic content of the output voltage of all higher-voltage cells. This command signal is compared with a voltage level corresponding to the sum of all voltage sources up to the HB_{i-1} ($\sigma_{max,i-1}$). In the same way as for HB_h, the output voltage of this cell is synthesized from a comparison of these two signals.

Finally, the command signal of HB_1 (lowest power inverter) is compared with a high-frequency triangle carrier signal, resulting in a high-frequency output voltage. Therefore, the output voltage harmonics will be concentrated around the frequencies that are multiples of the switching frequency of the inverter with the lowest DC voltage source. Consequently, the spectral response of the output voltage depends on the switching frequency of the lowest power inverter, while the power processing depends on the inverter with the highest DC voltage source.

However, with the hybrid modulation strategy, a voltage waveform must be synthesized to modulate at high frequency among all adjacent voltage steps. Only the lower-voltage HB can switch at high frequency, so the DC voltages must satisfy the following equation:

$$V_{dci^*} \leq 2 \sum_{k=1}^{i-1} V_{dck^*}, \quad j = 2, 3, \dots, h \qquad (9.32)$$

where * indicates the normalized value. Therefore, the hybrid modulation strategy can be applied in binary hybrid multilevel inverters and quasi-linear multilevel inverters. The relationship of DC voltages of the THMI is shown in Equation (9.15), so the THMI cannot use the hybrid modulation strategy.

9.2.3.4 Subharmonic PWM Strategies

Subharmonic PWM strategies for multilevel inverters employ extensions of carrier-based techniques used for conventional inverters. It has been reported that the spectral performance of a five-level waveform can be significantly improved by employing alternative dispositions and phase shifts in the carrier signals. This concept can be extended to a nine-level case with the available options for polarity and phase variation. A representative subharmonic PWM waveform with the nine-level phase leg voltage is shown in Figure 9.14.

If a two-HB THMI is used to synthesize the nine-level phase leg voltage as shown in Figure 9.14b, the higher-voltage HBs will switch at high frequencies, since the output voltage varies between E and $2E$ or $-E$ and $-2E$ continually at a certain interval. In THMI, it is not appropriate for the higher-voltage HBs to switch at high frequency. Therefore, subharmonic PWM is not applicable in THMI.

9.2.3.5 Simple Modulation Strategy

In the simple modulation strategy, the switching pattern is determined by comparing a reference signal with stages and then choosing the stages

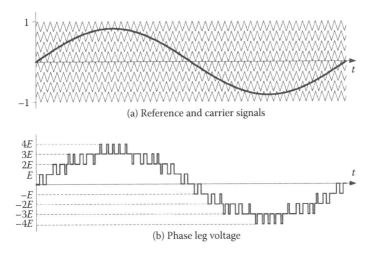

(a) Reference and carrier signals

(b) Phase leg voltage

FIGURE 9.14
Representative waveforms for subharmonic PWM waveform with carrier polarity variation.

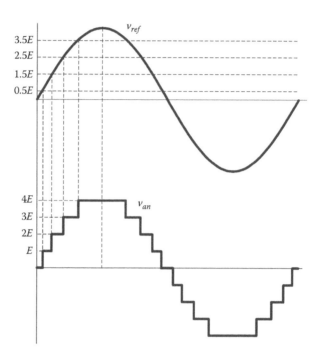

FIGURE 9.15
Illustration of the simple modulation strategy.

closest to the reference signal. Figure 9.15 shows the simple modulation strategy with the nine-level output voltage.

The advantages of this strategy are a simple control algorithm, high flexibility, and dynamic response. The disadvantage is that the amplitudes of lower-order harmonic components are relatively higher. The THMI can generate the greatest voltage levels among all multilevel inverters using the same number of components. If the number of voltage levels is high enough, the lower-order harmonic components of output voltages will be very small with the simple modulation strategy. For example, in the case of a four-HB THMI that can generate an 81-level voltage, with the simple modulation strategy, the amplitude of each lower-order harmonic component of the output voltage is less than 0.9% of the amplitude of the fundamental component and THD of output voltage is less than 2%.

Several modulation strategies have been investigated. With the hybrid modulation strategy and modulation strategies working with high switching frequencies, such as the subharmonic PWM strategy, a voltage waveform must be synthesized to modulate at high frequency among all adjacent voltage steps. However, in THMI, it cannot be achieved when only the lowest-voltage HBs switch at high frequency, which can be derived from Equations (9.15) and (9.32). In other words, if a voltage can be synthesized to modulate at high frequency in THMI, the higher-voltage HBs must switch at high frequency.

One of the most important advantages of THMI is that higher-voltage HBs can switch at lower frequency. Therefore, higher-frequency switching of higher-voltage HBs not only is unacceptable in high-power applications but also violates the main advantage of THMI. Therefore, the hybrid modulation strategy and other modulation strategies working with high switching frequencies are not applicable in THMI.

The low-frequency modulation strategies such as step modulation strategy and virtual stage modulation strategy are suitable in THMI. In virtual stage modulation, an additional constraint, such as Equation (9.31) for two-HB THMI, must be added to ensure the higher HBs switch at a lower frequency. Additionally, the simple modulation strategy can be used in the THMIs that can generate many voltage levels.

At the same time, for the THMIs that can generate many voltage levels, the space vector modulation can achieve a very good linearity between the modulation index and the fundamental component of load voltage and eliminate common-mode voltages.

9.2.4 Regenerative Power

The DC sources of the THMI can be batteries or bridge rectifiers. Large reverse current for a long time will damage batteries. Diode bridge rectifiers cannot permit reverse power. Controlled bridge rectifiers can transmit energy to supplies. However, compared to simple diode bridge rectifiers, controlled bridge rectifiers are much more complex and costly because of complex control circuits and the higher price of controlled semiconductors.

9.2.4.1 Analysis of DC Bus Power Injection

The switching function is involved in the analysis of DC bus power injection. The switching function, F, is shown in Table 9.1. The relationship between output voltage of a HB, v_H, and the DC link voltage of the HB, V_{dc}, can be written as Equation (9.33). The relationship between i_{dc} (current flowing through the DC bus) and i_{an} (output current of the THMI) can be also derived as Equation (9.34).

$$v_H = F \cdot V_{dc} \tag{9.33}$$

$$i_{dc} = F \cdot i_{an} \tag{9.34}$$

Only the fundamental component of output current of the THMI is considered since high-frequency harmonic components do not generate average power. So i_{an} can be expressed as

$$i_{an} = I_{an} \cdot \sin(\omega t + \varphi) \tag{9.35}$$

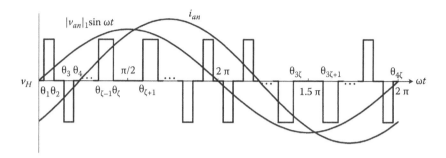

FIGURE 9.16
General waveform of DC bus voltage and THMI output current.

where I_{an} is the amplitude of i_{an} and φ is the angle of power factor. General waveforms of v_H and i_{an} are shown in Figure 9.16. The average DC power that supplies the HB over a period can be calculated as

$$P_{dc} = \frac{1}{T}\int_0^T V_{dc} \cdot i_{dc}\, dt \qquad (9.36)$$

where T is the period of i_{an}. From Equations (9.34) and (9.36), we can get

$$P_{dc} = \frac{1}{T}\int_0^T V_{dc} \cdot F \cdot i_{an} dt = \frac{1}{T}\int_0^T v_H \cdot i_{an}\, dt \qquad (9.37)$$

The relationship between switching angles in Figure 9.16 can be expressed as

$$\theta_i = \begin{cases} \pi + \theta_{i-2\varsigma} & i = (2\varsigma+1)\cdots 4\varsigma \\ \pi - \theta_{2\varsigma+1-i} & i = (\varsigma+1)\cdots 2\varsigma \end{cases} \qquad (9.38)$$

From Equation (9.38), v_H has following characteristic:

$$v_H(\pi - \omega t) = v_H(\omega t) \qquad (9.39)$$

$$v_H(\omega t + \pi) = -v_H(\omega t) \qquad (9.40)$$

From Equation (9.35), we obtain

$$i_{an}(\omega t + \pi) = -i_{an}(\omega t) \qquad (9.41)$$

From Equations (9.40) and (9.41), the average DC power can be calculated over a half period as

$$P_{dc} = \frac{2}{T} \int_0^{\frac{T}{2}} v_H \cdot i_{an}\, dt \tag{9.42}$$

Suppose P_i is the power generated by the voltage pulse from θ_i/ω to θ_{i+1}/ω and the corresponding voltage pulse ranges from $\theta_{2\varsigma-i}/\omega$ to $\theta_{2\varsigma+1-i}/\omega$. P_i can be expressed as:

$$P_i = \frac{(-1)^n \omega}{\pi} \left(\int_{\theta_i/\omega}^{\theta_{i+1}/\omega} V_{dc} \cdot I_{an} \cdot \sin(\omega t + \varphi)dt + \int_{\theta_{2\varsigma-i}/\omega}^{\theta_{2\varsigma+1-i}/\omega} V_{dc} \cdot I_{an} \cdot \sin(\omega t + \varphi)dt \right) \tag{9.43}$$

where $i = 2n - 1$ and n is a natural number. From Equations (9.38) and (9.43), P_i is expressed as

$$P_i = \frac{(-1)^n}{\pi} V_{dc} \cdot I_{an} \cdot 2 \cdot \cos(\varphi) \cdot (\cos(\theta_i) - \cos(\theta_{i+1})) \tag{9.44}$$

Thus, the average DC power of the HB can be expressed as

$$P_{dc} = \frac{2}{\pi} V_{dc} \cdot I_{an} \cdot \cos(\varphi) \cdot \sum (\cos(\theta_{4n-3}) - \cos(\theta_{4n-2}) - \cos(\theta_{4n-1}) + \cos(\theta_{4n})) \tag{9.45}$$

In Equation (9.45), if θ_j is greater than $\pi/2$, θ_j will be set as $\pi/2$.

In general, the power factor angle φ is from $-\pi/2$ to $\pi/2$, so $\cos(\varphi)$ is greater than zero. V_{dc} and I_{an} are positive. Thus, we can conclude from Equation (9.46) that the power of the DC bus is negative if

$$\sum (\cos(\theta_{4n+1}) - \cos(\theta_{4n+2}) - \cos(\theta_{4n+3}) + \cos(\theta_{4n+4})) < 0 \tag{9.46}$$

Negative power of the DC bus means regenerative power.

9.2.4.2 Regenerative Power in THMI

Regenerative power may occur in lower-voltage HBs of THMI. Take the example of a two-HB THMI. If the step modulation strategy is applied, the restrictions that are added to Equation (9.30) to ensure that the power of DC buses is always positive are shown in Table 9.8. With these restrictions,

TABLE 9.8

Additional Restriction to Avoid Regenerative
Power of DC Buses in Step Modulation

σ	Restriction
1	$\cos(\theta_1) > 0$
2	$\cos(\theta_1) - 2\cos(\theta_2) > 0$
3	$\cos(\theta_1) - 2\cos(\theta_2) + \cos(\theta_3) > 0$
4	$\cos(\theta_1) - 2\cos(\theta_2) + \cos(\theta_3) + \cos(\theta_4) > 0$

ranges of relative modulation index are calculated as shown in Equation (9.9). The range of the relative modulation index decreases greatly when σ is two or three compared with Table 9.9. The range of the modulation index is not continuous, as shown in the last column of Table 9.9. The regenerative power will occur in the lower-voltage HB when M varies from 0.51 to 0.55 or from 0.33 to 0.44.

Consider the virtual stage modulation strategy is used in a two-HB THMI. In Table 9.9, two cases are analyzed. One is the four-level positive/negative output voltage with two virtual stages, and the other is the five-level positive/negative output voltage with one virtual stage. Only the DC bus of the lower-voltage HB can have regenerative power. For the first case, the restriction that ensures positive power can be written as

$$\cos(\theta_{p1}) - 2\cos(\theta_{p2}) + \cos(\theta_{p3}) + \cos(\theta_{p4}) + \cos(\theta_{p5}) + \cos(\theta_{p6}) - \cos(\theta_{n1})$$
$$- \cos(\theta_{n2}) > 0 \tag{9.47}$$

In the second case, the restriction can be expressed as

$$\cos(\theta_{p1}) - 2\cos(\theta_{p2}) + \cos(\theta_{p3}) + \cos(\theta_{p4}) + \cos(\theta_{p5}) - \cos(\theta_{n1}) > 0 \tag{9.48}$$

With these restrictions, the range of the relative modulation index decreases as shown in Table 9.10. The regenerative power will occur in the lower-voltage HB when M varies from 0.53 to 0.62.

TABLE 9.9

Range of Modulation Index with the Step Modulation in a
two-HB THMI (Avoid Regenerative Power of DC Buses)

σ	MR (min)	MR (max)	M (min)	M (max)	Range of MA
1	0	1	0	0.25	0–0.15
2	0.3	0.66	0.15	0.33	0.15–0.33
3	0.59	0.68	0.44	0.51	0.44–0.51
4	0.55	0.86	0.55	0.86	0.55–0.86
4*	0.56	0.94	0.56	0.94	0.86–0.94

TABLE 9.10

Range of Modulation Index Range with the Virtual Stage
Modulation Strategy in a Two-HB THMI (Avoiding Regenerative
Power of DC Bus)

σ	β	MR (min)	MR (max)	M (min)	M (max)	Range of M
3	1	0.62	0.71	0.46	0.53	0.46–0.53
4	2	0.62	0.92	0.62	0.92	0.62–0.92

9.2.4.3 Method to Avoid Regenerative Power

In the last section, the regenerative power that lower-voltage HBs can generate is discussed. In this section, the methods that are used to solve this problem will be introduced. A method is proposed in which the DC links of lower-voltage HBs are supplied by the low-power, isolated power sources fed by a common power supply from the highest-voltage HB. These power sources are bidirectional, and a bidirectional DC-DC power supply is used for this purpose. It is also possible to use independent output transformers with a common DC supply, as shown in Figure 9.17. A variation of this configuration was used in a 16 2/3 Hz substation for railroads in Bremen (Germany). In the system described here, the transforms are smaller for lower-voltage HBs because the voltages are scaled in powers of three. Besides, the switching frequencies of transformers connected with lower-voltage HBs are lower. Then the transforms connected with lower-voltage HBs become smaller for two reasons: voltage and switching frequency.

The foregoing two methods of solving the problem of regenerative power use additional equipment such as bidirectional DC/DC converters or output transformers, which increase the cost of the inverter system and power losses. A new method is presented to avoid regenerative power that does not use additional devices. Regenerative power is eliminated by avoiding outputting several null voltage levels, which will be explained by an example of a four-HB THMI in the following.

The average power of the DC bus of an HB can be expressed as Equation (9.45). In general, the power factor angle φ varies from $-\pi/2$ to $\pi/2$, so $\cos(\varphi)$ is greater than zero. V_{dc} and I_{an} are positive. Therefore, from Equation (9.45) and Figure 9.15, we can conclude that the reason for the regenerative power is the negative segments of v_H when the fundamental components of v_{an} are positive or the positive segments of v_H when the fundamental components of v_{an} are negative. The segments of v_H resulting in the regenerative power of the HB are called regenerative segments. The basic idea of eliminating regenerative power is to avoid output several levels of v_{an}, which will cause regenerative segments in HBs. Table 9.11 shows the voltage levels of v_{an} that cause regenerative segments of HBs in the case of a four-HB THMI. The voltage levels of v_{an} that are not selected for output are called null voltage levels. Table 9.11 also shows the priority

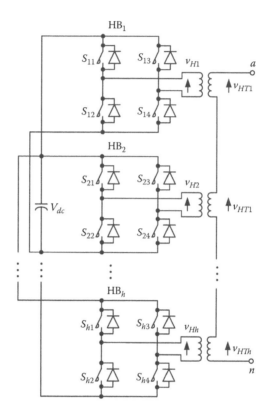

FIGURE 9.17
THMI with output transformers.

of null voltage levels. For example, if the regenerative power occurs in the DC link of the HB_1, the voltage levels (14) and (–14) are selected as null voltage levels first. If the regenerative power still occurs, levels (17) and (–17) are also selected as null voltage levels. With the priority shown in Table 9.11, the null voltage levels distribute as evenly as possible, which results in better power quality.

Figure 9.18 shows the flowchart of the algorithm that stabilizes the DC link voltage of an HB. V_{dc} is the DC link voltage of an HB. $V_{dc,normal}$ is the normal DC link voltage. $V_{dc,last}$ is the DC link voltage in the previous sampling. N_{null}

TABLE 9.11

Voltage Levels of v_{an} that Cause Regenerative Segments of HBs

HB_1	±14, ±17, ±32, ±23, ±5, ±20, ±38, ±29, ±26, ±11, ±2, ±8, ±35
HB_2	±14, ±32, ±23, ±5, ±15, ±34, ±25, ±7, ±16, ±6, ±24, ±33
HB_3	±14, ±17, ±15, ±20, ±19, ±16, ±21, ±18, ±22

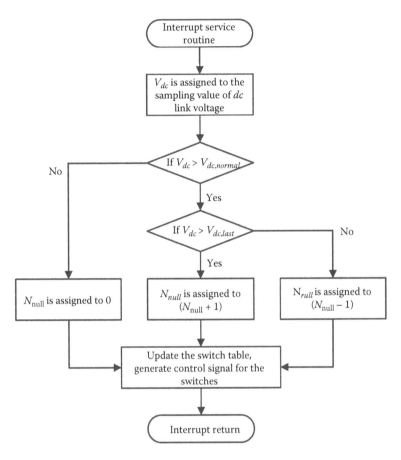

FIGURE 9.18
Flowchart of the algorithm to stabilize DC link voltages.

is the number of null voltage levels. In the switch table, the voltage levels are set as null or not based on N_{null} and Table 9.11.

With a lower modulation index, the power quality that the THMI outputs is a bit poorer with the proposed control scheme because more null voltage levels are not dedicated to the output voltage of the THMI. In the case of the four-HB THMI, with up to 81-level output voltage of the THMI, the simple modulation strategy is suitable for the THMI. If the simple modulation strategy is used and the new method is applied to eliminate the effect of regenerative power, the relationship between the modulation index and the THD is shown in Figure 9.19.

9.2.4.4 Summary of Regenerative Power in THMI

The topology of THMI has the distinct advantage of fewer components used compared with other topologies of multilevel inverters, but the THMI also has

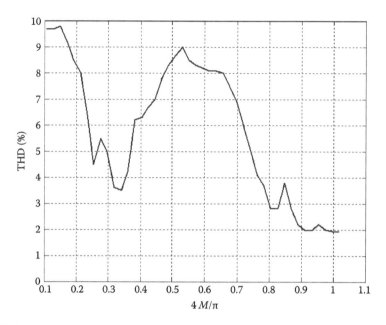

FIGURE 9.19
Relationship between the modulation index and THD.

the notable disadvantage that the power of the lower-voltage HBs can be regenerative with a lower modulation index. If the THMI feeds an RL or RC load and simple diode bridge rectifiers are used as DC sources, the regenerative power will cause an increase of the DC capacitor voltages, which will damage the devices.

Therefore, basically, the THMI is suitable for two applications. The first one is the application of reactive power compensation. The average power of the DC link of an HB is zero when the power factor angle is zero as shown in Equation (9.45), so the problem of regenerative power is avoided. The second one is the application in which the inverter always runs with a higher modulation index. From Table 9.9, we can see that the two-HB THMI runs with step modulation without the problem of regenerative power when M varies from 0.56 to 0.94. From Table 9.10, we can see that the two-HB THMI runs with virtual stage modulation without the problem of regenerative power when M varies from 0.62 to 0.92.

However, the inverter is required to work at any modulation index for active load in most cases. Two methods have been presented to solve the regenerative power problem. The first one uses bidirectional DC/DC converters, and the second one uses additional output transformers. A new method to solve the regenerative power is presented as a cost-effective solution because it does not use additional equipment. The DC capacitor voltages of lower-voltage HBs are kept stable by the new method. The trade-off is that power quality will decrease a bit with the lower modulation index.

9.3 Experimental Results

Some experimental results are shown here to help readers gain a deeper understanding.

9.3.1 Experiment to Verify Step Modulation and Virtual Stage Modulation

The performance of the step modulation strategy and virtual stage modulation strategy has been verified by the experiment of a single two-HB THMI. In the control circuit, a TMS320F240 DSP is used as the main processor, which provides the gate logic signals. In an HB, four MOSFETs, IRF540, are used as the main switches, which are connected in a full-bridge configuration. The load is a 23.2 Ω resistor. The total ratio of voltage measure is 1:2. The frequency spectrums are analyzed by the FFT (fast Fourier transform) function of the oscilloscope. The scale of the Y-axis of the frequency spectrum is 5dBV/div, and the reference level is 5 dBV.

The switching pattern of the step modulation technique is programmed and is loaded into the DSP. In the step modulation strategy, when there are nine output voltage levels and M is 0.83, the switching angles are 0.14778, 0.32325, 0.57376, and 0.99696. The THMI output voltage is shown in Figure 9.20. The frequency is 50 Hz, and the step voltage is about 5 V. The

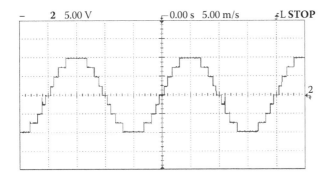

FIGURE 9.20
Output voltage of the THMI with the step modulation $M = 0.83$ (10 V/div).

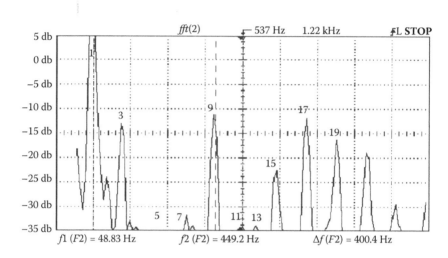

FIGURE 9.21
Frequency spectrum with the step modulation $M = 0.83$.

frequency spectrum is shown in Figure 9.21. The 5th, 7th, and 11th harmonics are less than 0.028 V (-37dB $\times$ 2 V), which means they are nearly eliminated.

When there are seven output voltage levels and M is 0.49, the switching angles are 0.44717, 0.9097, and 1.1215. The output voltage of the THMI is shown in Figure 9.22, and the frequency spectrum is shown in Figure 9.23. The 5th and 7th harmonics are less than 0.02 V (-40dB $\times$ 2 V), which means they are nearly eliminated.

When there are five output voltage levels and M is 0.32, the switching angles are 0.51847 and 1.1468. The output voltage of the THMI is shown in Figure 9.24, and the frequency spectrum is shown in Figure 9.25. The 5th

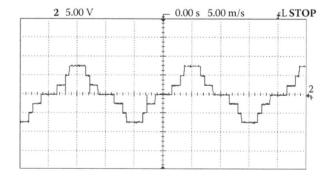

FIGURE 9.22
Output voltage of the THMI with the step modulation $M = 0.49$ (10 V/div).

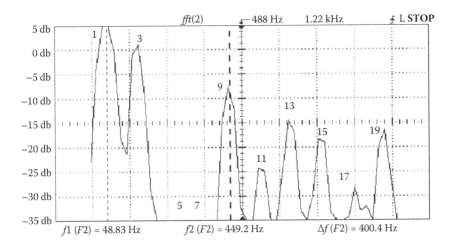

FIGURE 9.23
Frequency spectrum with the step modulation $M = 0.49$.

harmonics are less than 0.02 V (–40dB × 2 V), which means they are nearly eliminated.

The switching pattern of the modified virtual stage modulation technique is programmed and is loaded to the DSP. In virtual stage modulation, when there are nine output voltage levels and M is 0.83, the switching angles are 0.13177, 0.33186, 0.52855, 0.6202, 0.91294, 1.0423, 0.57124, and 0.96573. The output voltage of the THMI is shown in Figure 9.26, and the frequency spectrum is shown in Figure 9.27. The 5th, 7th, 11th, 13th, 17th, 19th, and 23rd harmonics are less than 0.035 V (–35dB × 2V), which means they are nearly eliminated.

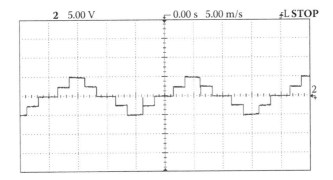

FIGURE 9.24
Output voltage of the THMI with the step modulation $M = 0.32$ (10 V/div).

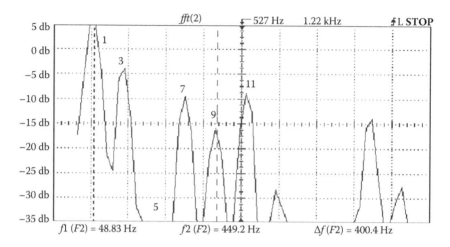

FIGURE 9.25
Frequency spectrum with the step modulation $M = 0.32$.

When there are seven output voltage levels and M is 0.49, the switching angles are 0.40549, 0.88038, 1.1497, 1.5318, and 1.5082. The output voltage of the THMI is shown in Figure 9.28, and the frequency spectrum is shown in Figure 9.29. The 5th, 7th, 11th, and 13th harmonics are less than 0.02 V (-40 dB $\times$ 2V), which means they are nearly eliminated.

9.3.2 Experiment to Verify New Method to Eliminate Regenerative Power

The performance of the methods to eliminate the effect of regenerative power by avoiding outputting the null voltage levels is verified by the

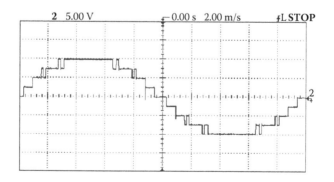

FIGURE 9.26
Output voltage of the THMI with the virtual stage modulation $M = 0.83$ (10 V/div).

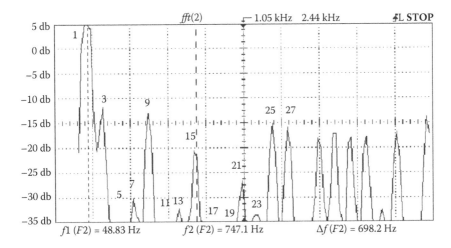

FIGURE 9.27
Frequency spectrum with the virtual stage modulation $M = 0.83$.

experiment of a 4-HB THMI, in which diode bridge rectifiers are used as the DC sources of HBs. The step voltage is 5.9 V. The frequency of output voltage is set at 50 Hz, and the sampling frequency is set at 10 kHz. The output voltage of the inverter has up to 81 levels, so the simple modulation strategy as shown in section 9.2. The control algorithm to stabilize the DC link voltages is shown in Figure 9.26. A TMS320F240 DSP-controlled card is used to control the inverter. The configuration of the experimental system is shown in Figure 9.30.

Figure 9.31 shows the waveform of the output voltage of the 4-HB THMI with simple modulation strategy when the modulation index is 0.79. The power quality is good due to of the large number of voltage levels.

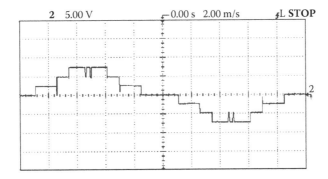

FIGURE 9.28
Output voltage of the THMI with the virtual stage modulation $M = 0.49$ (10 V/div).

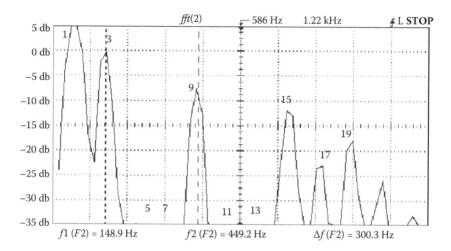

FIGURE 9.29
Frequency spectrum with the virtual stage modulation $M = 0.49$.

Figure 9.32 shows the output voltage waveform with some null voltage levels when the modulation index is 0.7. From the enlarged figure, we can observe that some voltage levels are not generated. Moreover, the step voltages are kept nearly the same, which means that the voltages of DC capacitors are kept stable. Figure 9.33 shows the worst case when the modulation index is 0.53. In this case, null voltage levels include ±5, ±14, ±15, ±16, ±17, ±19, ±20, ±21, ±23, ±32, and ±34.

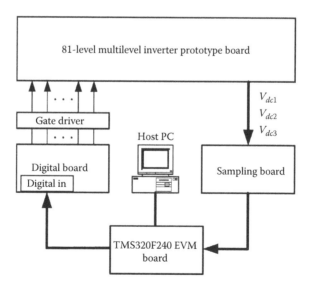

FIGURE 9.30
General representation of experimental test system.

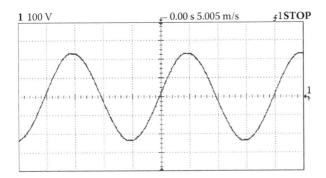

FIGURE 9.31
Waveform of output voltage of the inverter with the simple modulation strategy $M = 0.79$ (100 V/div), frequency = 50 Hz, THD = 1.94%.

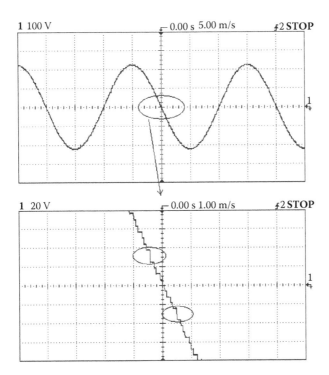

FIGURE 9.32
Waveform of output voltage of the inverter $M = 0.7$ (100 V/div).

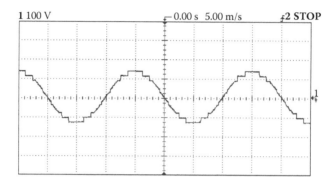

FIGURE 9.33
Waveform of output voltage of the inverter $M = 0.42$ (100 V/div).

9.4 Trinary Hybrid 81-Level Multilevel Inverter

A trinary hybrid 81-level multilevel inverter for motor drive with zero common-mode voltage was designed. Figure 9.34 shows the power circuit topology of the trinary hybrid multilevel inverter for a motor drive. Bidirectional DC/DC converters connected to the DC link of H-bridges feed HBs. To get the maximum output voltage levels of the inverter, the ratio of DC link voltages is arranged as 1:3:9:27, so the inverter can output 81 voltage levels in each phase. With four HBs per phase, however, a cascade multilevel inverter [2–4] can output only 9 voltage levels each phase, and a binary hybrid multilevel inverter [5] can output only 31 voltage levels each phase. The more output voltage levels a multilevel inverter has, the more nearly sinusoidal the synthesized waveform. Thus, with the trinary hybrid topology, total harmonic distortion (THD) can be greatly reduced.

Three phases of the inverter are controlled separately, and the operating principle of each phase is identical. In the following, the A-phase of the inverter is analyzed. HB_{ak} represents the kth HB in the A-phase leg of the inverter, as shown in Figure 9.35. $v_{H,ak}$ and $v_{C,ak}$ represent the output voltage and the DC link voltage of the HB_{ak}, respectively. A switching function, F_{ak}, is used to relate $v_{H,ak}$ and $v_{C,ak}$ as

$$v_{H,ak} = F_{ak} \cdot v_{C,ak} \quad (k = 1..4) \tag{9.49}$$

The value of F_{ak} can be 1, –1, or 0. For the value 1, the upper switch of the left leg and the lower switch of the right leg in an HB need to be turned on. For the value –1, the lower switch of the left leg and the upper switch of the right leg in an HB need to be turned on. For the value 0, upper switches of both

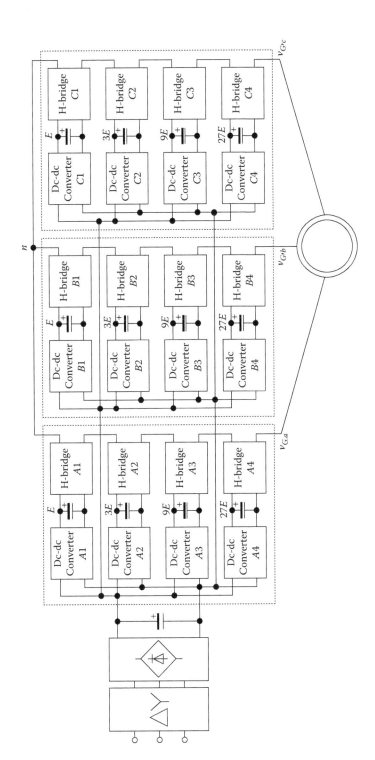

FIGURE 9.34
Power circuit topology of the trinary hybrid 81-level inverter for motor drives.

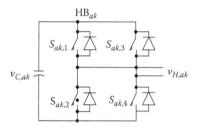

FIGURE 9.35
An H-bridge.

legs or lower switches of both legs need to be turned on. The A-phase voltage of the inverter, $v_{G,a}$, is represented as

$$v_{G,a} = \sum_{1}^{4}(F_{ak} \cdot v_{C,ak})$$

(9.50)

Suppose l is the ordinal of expected voltage level that the inverter outputs. If l is positive or zero, the inverter outputs the positive lth voltage level. If l is negative, the inverter outputs the negative $(-l)$th voltage level. The value of F_{ak} can be determined by

$$F_{a4} = B_u(\text{ABS}(\ell) - 13) \cdot \text{ABS}(\ell)/\ell \qquad \ell_3 = \ell - F_{a4} \cdot 27$$
$$F_{a3} = B_u(\text{ABS}(\ell_3) - 4) \cdot \text{ABS}(\ell_3)/\ell_3 \qquad \ell_2 = \ell_3 - F_{a3} \cdot 9$$
$$F_{a2} = B_u(\text{ABS}(\ell_2) - 1) \cdot \text{ABS}(\ell_2)/\ell_2 \qquad \ell_1 = \ell_2 - F_{a2} \cdot 3$$
$$F_{a1} = B_u(\text{ABS}(\ell_1)) \cdot \text{ABS}(\ell_1)/\ell_1$$

(9.51)

where ABS is the function of absolute value and B_u is defined as

$$B_u(\tau) = \begin{cases} 1 & \tau > 0 \\ 0 & \tau \leq 0 \end{cases}$$

(9.52)

From Equation (9.51), we can get the relationship between the output voltage of a phase leg and the values of switching functions of HBs in a phase leg.

9.4.1 Space Vector Modulation

$v_{G,a}$, $v_{G,b}$, and $v_{G,c}$ are the voltages of terminals a, b, and c of the inverter with respect to the neutral n [9]. Three-phase inverter output voltages can be represented by a space vector in the x-y plane using the following transformation:

$$v = v_x + jv_y = \frac{2}{3}(v_{G,a} + \alpha v_{G,b} + \alpha^2 v_{G,c})$$

(9.53)

where

$$\alpha = -\frac{1}{2} + j\frac{\sqrt{3}}{2} \qquad (9.54)$$

Equation (9.53) can be expressed as a function of their real and imaginary components:

$$v_x = \frac{1}{3}(2v_{G,a} - v_{G,b} - v_{G,c}) \qquad (9.55)$$

$$v_y = \frac{1}{\sqrt{3}}(v_{G,b} - v_{G,c}) \qquad (9.56)$$

The number of different voltage vectors is represented as

$$N_v = 2N_l - 1 + \sum_{i=1}^{2(N_l-1)} 2i \qquad (9.57)$$

where N_l is the number of voltage levels. Each phase can generate 81 different voltages, so totally 19,411 different voltage vectors can be generated as shown in Figure 9.36.

The common mode voltage is defined as

$$v_{cm} = \frac{1}{3}(v_{G,a} + v_{G,b} + v_{G,c}) \qquad (9.58)$$

Considering this definition, we can find vectors generated by three phase voltages that produce zero common mode voltage as shown in Figure 9.37. The use of only vectors that generate zero common mode voltages to the load reduces the density of vectors available to be applied. The number of different authorized licensed voltage vectors with zero common mode voltage is represented as

$$N_{vz} = \frac{3N_l^2 + 1}{4} \qquad (9.59)$$

Therefore, there are still 4,921 different voltage vectors available.

In Figure 9.38, the nearest voltage vector with respect to the reference vector v_{ref} is delivered. The following algorithm is used to select the appropriate vector based on the information of the reference vector.

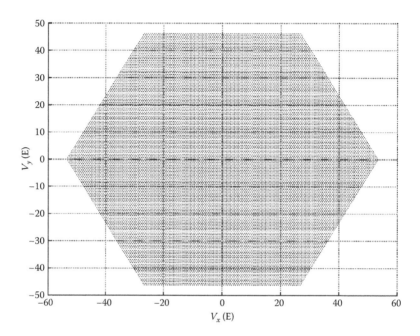

FIGURE 9.36
Voltage vectors of a three-phase 81-level inverter.

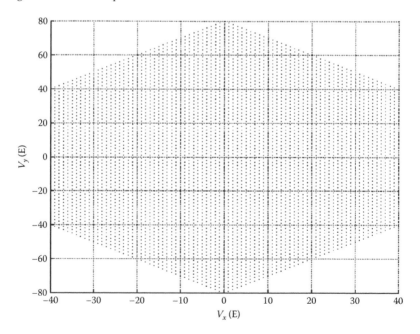

FIGURE 9.37
Voltage vectors of a three-phase 81-level inverter with zero common mode voltage.

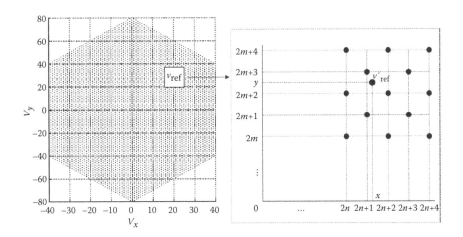

FIGURE 9.38

Normalized voltage vectors of a three-phase 81-level inverter with zero common mode voltage.

Step 1. Normalize the reference vector $v_{ref} = v_{xref} + j v_{yref}$

$$v'_{ref} = \frac{1}{E} v_{xref} + j \frac{\sqrt{3}}{E} v_{yref} = x + jy \qquad (9.60)$$

Step 2. Normalize the candidate space vector with the transformation in Equation (9.60), converting them into integer values. After conversion, the space vectors with zero common mode voltage are shown in Figure 9.37. The addition of the x-axis value and y-axis value of each space vector with zero common mode voltage is an even number.

Step 3. v'_{ref} will lie in one of the rectangles defined by two normalized candidate space vectors as shown in the right part of Figure 9.38. The rectangle is identified by the values of left bottom point of the rectangle. $v'_{ref}(x, y)$ lies in the rectangle $(floor(x), floor(y))$, where $floor(\alpha)$ is the function that rounds the elements of α to the nearest integer that is less than or equal to α. In the rectangle $(floor(x), floor(y))$, there are two normalized voltage vectors, $(floor(x), floor(y))$ and $(floor(x)+1, floor(y)+1)$, if the addition of $floor(x)$ and $floor(y)$ is even. Two vectors are $(floor(x)+1, y)$ and $(x, floor(y)+1)$ if the addition of $floor(x)$ and $floor(y)$ is odd. Suppose the reference vector, $v'_{ref}(x, y)$ lies in the rectangle with two normalized voltage vectors, v_1 and v_2. The nearest vector is selected by comparing the distances of each candidate vector, v_1 and v_2, with respect to v'_{ref} using the following equations:

$$d_1 = \sqrt{(3(x - Re(v_1))^2 - (y - Im((v_1))^2} \qquad (9.61)$$

$$d_2 = \sqrt{(3(x - Re(v_2))^2 - (y - Im(v_2))^2} \qquad (9.62)$$

The selection is done by

$$if\ d_1 < d_2\ then\ v_{sel} = v_1$$
$$else\ v_{sel} = v_2$$

(9.63)

Step 4. Three-phase output voltages with zero common mode voltage are generated by an inverse transformation for v_{sel} as

$$v_{G,a} = round(\text{Re}(v_{sel}))$$

$$v_{G,b} = v_{G,a} + \frac{\text{Im}(v_{sel}) - 3\,\text{Re}(v_{sel})}{2}$$

(9.64)

$$v_{G,c} = v_{G,a} - \frac{\text{Im}(v_{sel}) + 3\,\text{Re}(v_{sel})}{2}$$

9.4.2 DC Sources of H-Bridges

There are three reasons to set DC sources of HBs as bidirectional DC/DC converters in the proposed topology. The first reason is that the bidirectional DC/DC converter can transfer the regenerative power from the HB to the rectifier. In an HB, the output voltage is v_H, and the current flowing through the HB is i_H, as shown in Figure 9.39.

Only the fundamental component of output current of the inverter is considered since high-frequency harmonic components do not generate average power. Thus, the average power flowing through the DC link of the HB, $P_{H,dc}$, can be expressed as

$$P_{H,dc} = \frac{2}{\pi} v_C I_H \cos\varphi \sum (\cos(\theta_{4n-3}) - \cos(\theta_{4n-2}) - \cos(\theta_{4n-1}) - \cos(\theta_{4n}))\ n = 1,2,..$$

(9.65)

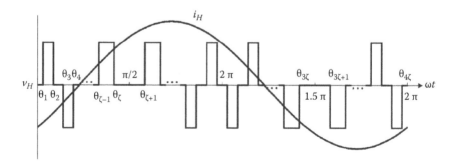

FIGURE 9.39
General waveform of output voltage and current of an HB.

where v_C is the DC link voltage of the HB and I_H is the amplitude of i_H. φ is the power factor angle for fundamental components of v_H and i_H. In Equation (9.65), if θ_j is greater than $\pi/2$, θ_j it will be set as $\pi/2$. In general, the power factor angle φ varies from $-\pi/2$ to $\pi/2$, so $\cos(\varphi)$ is greater than zero. v_C and I_H are positive. Thus, we can conclude from Equation (9.65) that the power of DC link is negative if

$$\sum(\cos(\theta_{4n-3}) - \cos(\theta_{4n-2}) - \cos(\theta_{4n-1}) - \cos(\theta_{4n})) < 0 \quad n = 1, 2. \quad (9.66)$$

When the inverter feeds the motor, the power of the DC link of the HB with the highest DC voltages is always positive. However, the power of the DC link of other HBs may be negative with a lower modulation index. Therefore, the bidirectional DC/DC converter is necessary here to transfer regenerative power of the DC link back to the rectifier to avoid the increase of the DC link voltage.

The second reason is that variation of the DC link voltage of an HB is required to be very small. For example, the variation of the DC link voltage of the HB with a DC link voltage of 27 E must be less than 0.5/27 = 0.019. Otherwise, the contribution of the HB with DC link voltage of E for the power quality will be negligible. DC/DC converters with high bandwidth close loop control can stabilize the DC link voltages of HBs.

The third reason is that transformers used in bidirectional converters are small, cost-effective, and high efficient. In other topologies of hybrid multilevel inverters for motor drives, the output ports of HBs are connected together by transformers. However, these low-frequency transformers are bulky and inefficient. Compared with configurations with low-frequency transformers, the efficiency of the DC/DC converter is higher. The efficiency of the DC/DC converter measured in the low power experiments is around 90%. In practical high-power application, that can reach 97% [6], which is much higher than that of the traditional configuration of low-frequency transformers and rectifiers.

Several topologies of bidirectional DC/DC converters were proposed [7,8]. The topology of the bidirectional DC/DC converter [8] is used in the proposed system shown in Figure 9.40. The transformer provides galvanic isolation between the input and output. The primary side of the converter is a half-bridge and is connected to the DC link of rectifier. All DC/DC converters share a diode rectifier as shown in Figure 9.40. The secondary side, connected to the DC link of the HB, forms a current-fed push–pull. The converter has two modes of operation. In the forward mode, the DC link of an HB is powered by the DC link of the rectifier. In the backward mode, the DC link of an HB provides the energy to the DC link of the rectifier.

The left part of Figure 9.41 shows the idealized waveforms in the forward mode: *Interval t_0-t_1*: Switch S_2 is off, and S_1 is on at time t_0. A voltage across the primary winding is $v_{Cr}/2$. The body diode of switch S_4, D_{S4}, is forward biased. The current flow through S_1, i_{S1}, contributes to the linearly increasing inductor current and the transformer primary magnetizing current. *Interval t_1-t_2*: Switch S_1 is turned off at time t_1, and S_2 remains on. No power is transferred to the secondary side during this dead time interval since there is zero

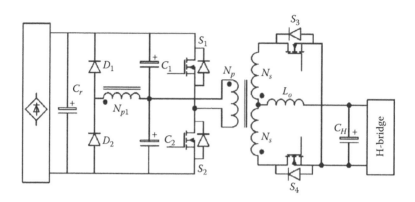

FIGURE 9.40
Bidirectional DC/DC converter.

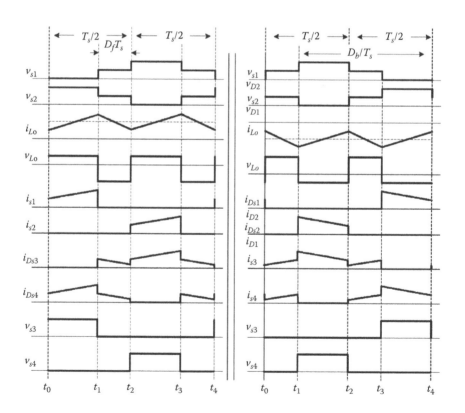

FIGURE 9.41
Waveforms of bidirectional DC/DC converter during the forward and backward modes.

voltage across the primary. The energy stored in L_o results in the freewheeling of the current i_{Lo} equally through the body diodes D_{S3} and D_{S4}. *Interval t_2-t_3*: Switch S_2 is turned on at time t_2, and S_1 remains off. The operation is similar to that during interval t_0-t_1, but now D_{S3} conducts and provides secondary-side rectification. Inductor current rises linearly again. *Interval t_3-t_4*: Switch S_2 is turned off at time t_2, and S_1 remains off. The operation is similar to that in the interval t_1-t_2. Figure 9.40 shows a balancing winding N_{p1} and two diodes D_1 and D_2 on the primary side of the half-bridge. They maintain the voltage at the junction of C_1 and C_2 to be equal to one half of the input voltage, and prevent a voltage shift to the transformer core. N_{p1} has the same number of turns as the winding N_p and is phased in series with it through the on time of S_1 and S_2.

In the backward mode, the switches S_3 and S_4 of the current-fed push-pull topology are driven at duty ratios greater than 0.5. The converter operation during this mode is shown in the right part of Figure 9.41. *Interval t_0-t_1*: Switch S_3 is turned on, and S_4 remains on at time t_0. N_S is subject to a short circuit, which causes the inductor L_o to store energy as the DC link voltage of the HB appears across it. i_{Lo} ramps up linearly and is shared equally by both S_3 and S_4. During this interval, C_1 and C_2 provide the output power. *Interval t_1-t_2*: Switch S_4 is turned off, and S_3 remains on at time t_1. The energy stored in the inductor during the previous interval is now transferred to the load through D_{S2} and D_1. Voltages across N_{p1} and N_p are identical due to their series phasing and equal number of turns. This allows simultaneous and equal charging of both C_1 and C_2 through D_1 and D_{S2}, respectively. *Interval t_2-t_3*: Switch S_4 is turned on, and S_3 remains on at time t_2. This interval is similar to the interval t_0-t_1. The duty ratio for S_3 is therefore greater than 0.5. *Interval t_3-t_4*: Switch S_3 is turned off, and S_4 remains on at time t_3. The stored energy of L_o is transferred to the primary side of the converter through S_4, D_{S1}, and D_2. The conduction of D_{S1} and D_2 results in equal charging of C_1 and C_2, respectively. Current mode control is used for both modes of converter operations. Small signal analysis for both modes under mode control is performed to generate the transfer functions to design and evaluate the control loop [8].

9.4.3 Motor Controller

The proposed multilevel inverter is used to feed an induction motor. Vector control technique is applied in the motor controller. Vector control implies independent control of flux current and torque current components of stator current through a coordinated change in the supply voltage amplitude, phase, and frequency. As the flux variation tends to be slow, constancy of flux should produce a fast torque current response and finally a fast speed (position) response.

The controller is shown in Figure 9.42, and the current decoupling network in the controller is shown in Figure 9.43. To simplify the current decoupling network, the rotor flux orientation is used in the current decoupling network. Once the reference d–q current $i_{da}{}^*$, $i_{qa}{}^*$, and flux orientation angle $\theta_{er} + \gamma_a{}^*$ are known,

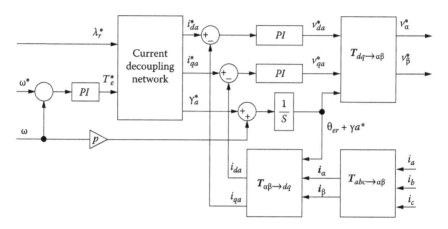

FIGURE 9.42
Motor controller.

the DC current controllers are used to translate these commands to v_{da}^* and v_{qa}^*, and use Park transformation to translate v_{da}^* and v_{qa}^* to v_α^* and v_β^*. The output signal of the motor controller, v_α^* and v_β^*, will be sent to the inverter controller to control the multilevel inverter to provide the appropriate voltages to feed the motor.

9.4.4 Simulation and Experimental Results

The performance of the 81-level trinary hybrid multilevel inverter for motor drives presented earlier has been verified by simulation. The simulation investigations were performed with MATLAB Simulink. The unit voltage of the multilevel inverter, E, is set as 10 V. The modulation index is defined as

$$m = \frac{\pi \, |v_{an}|_1}{4 \times 40E} \tag{9.67}$$

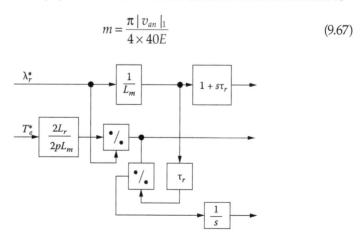

FIGURE 9.43
Current decoupling network.

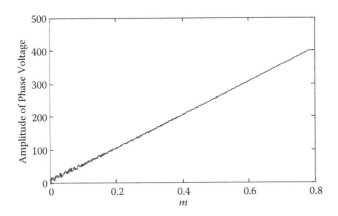

FIGURE 9.44
Amplitude of phase voltage versus modulation index.

where $|v_{an}|_1$ is the fundamental amplitude of output voltage. Based on the simulation results, the relationship between $|v_{an}|_1$ and the modulation index is shown in Figure 9.44. When the modulation index is very low, it does not have a very good linear relationship. However, due to the large number of voltage steps, the relationship becomes satisfied linearly with higher modulation indexes.

When the inverter drives an induction motor, a command of speed step change from 1430 to 715 rpm in 1ms. Figure 9.45 shows simulation results of speed, output voltage of the inverter, output current of the inverter, DC link voltages of HBs in the A-phase, and common-mode voltages. The speed has a rapid response. The common mode voltage is always zero except during the short transition time. Total harmonic distortion (THD) of output voltage is as low as 1%. Figure 9.46 shows the detailed waveforms of the output voltage of inverters. Figure 9.47 shows the simulation results of torque, output voltages, and output currents of the inverter when the reference torque step-changes from 1.29 to 7.74 Nm. The motor drive system also has a good dynamic response for the step change of torque.

To verify the performance of the proposed inverter experimentally, a hardware prototype has been built in the laboratory. The experimental setup of the proposed control system consists of a three-phase, 380 V, 50 Hz, 4 pole, 3-kW induction motor and power circuit using trinary hybrid multilevel inverter. The inverter and motor are controlled using TMS320F240 controller cards. The current mode controller of the DC/DC converters is implemented by UC 3846 and UCC 3804 for the forward mode and backward mode, respectively. Figures 9.48 and 9.49 show the waveforms of speed, phase current, phase voltage, and line-to-line voltage when the reference speed of the motor has a step change, which verifies the simulation results as shown in Figure 9.58. Figure 9.50 shows the detailed waveforms of phase voltage and common mode voltage. As shown in Figure 9.63, the phase voltage is synthesized by lots of stable step voltages, and the common mode voltage is almost zero.

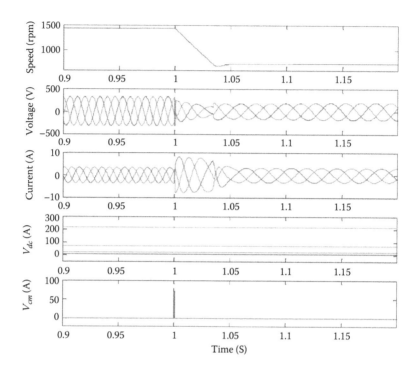

FIGURE 9.45
Simulation waveforms for a step change of speed.

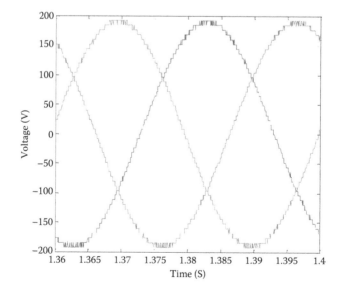

FIGURE 9.46
Simulation waveforms of output voltages of the inverter.

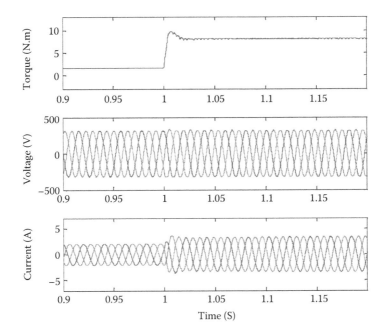

FIGURE 9.47
Simulation waveforms for a step change of torque (T from 1.29 Nm to 7.74 Nm).

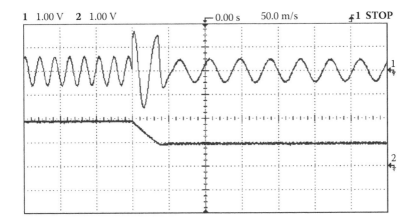

FIGURE 9.48
Experiment waveforms for a step change of speed. CH1: speed (750 rad/s/div); CH2: phase current (2 A/div).

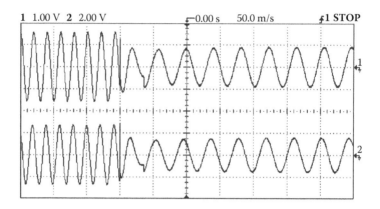

FIGURE 9.49
Experiment waveforms for a step change of speed. CH1: phase voltage (200 V/div); CH2: line-to-line voltage (400 V/div).

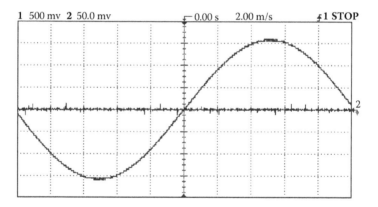

FIGURE 9.50
Experiment detailed waveforms. CH1: phase voltage (100 V/div); CH2: common mode voltage (20 V/div).

References

1. Liu, Y. and Luo, F. L. 2006. Multilevel inverter with the ability of self voltage balancing. *IEE Proc. Electric Power Applicat.*, pp. 105–115.
2. Hammond, P. W. 1997. New approach to enhance power quality for medium voltage ac drives. *IEEE Trans. Ind. Applicat.*, pp. 202–208.
3. Baker, R. H. and Bannister, L. H. 1975. Electric power converter. U.S. Patent 3 867 643.
4. Cengelci, E., Sulistijo, S. U., Woo, B. O., Enjeti, P., Teoderescu, R., and Blaabjerg, F. 1999. A new medium-voltage PWM inverter topology for adjustable-speed drives. *IEEE Trans. Ind. Applicat.*, pp. 628–637.
5. Manjrekar, M. D., Steimer, P. K., and Lipo, T. A. 2000. Hybrid multilevel power conversion system: a competitive solution for high-power applications. *IEEE Trans. Ind. Applicat.*, pp. 834–841.
6. Akagi, H. 2006. Medium-voltage power conversion systems in the next generation. *Proc. IEEE-IPEMC'2006*, pp. 23–30.
7. Inoue, S. and Akagi, H. 2007. A bidirectional isolated DC–DC converter as a core circuit of the next-generation medium-voltage power conversion system. *IEEE Trans. Power Electron.*, pp. 535–542.
8. Jain, M., Daniele, M., and Jain, P. K. 2000. A bidirectional DC–DC converter topology for low power application. *IEEE Trans. Power Electron.*, pp. 595-606.
9. Liu, Y. and Luo, F. L. 2008. Trinary hybrid 81-level multilevel inverter for motor drive with zero common-mode voltage. *IEEE Trans. Ind. Electron.*, pp. 1014–1021.

10

Laddered Multilevel DC/AC Inverters Used in Solar Panel Energy Systems

Multilevel DC/AC inverters have various structures and many advantages. Unfortunately, most existing inverters contain too many components (independent and floating batteries and sources, diodes, capacitors, and switches) making production cost high and conversion efficiency low. We introduced four kinds of inverters: laddered multilevel DC/AC inverters, super-lift modulated inverters, switched capacitor multilevel DC/AC inverters, and switched inductor multilevel DC/AC inverters in Chapters 10–13 that are new developments in this area. Their simple structures and straightforward operations are completely different from existing inverters. They have been successfully applied in solar panel energy systems [1]. These inverters will attract worldwide attention and be applied in other renewable energy systems.

10.1 Introduction

Comparing with PWM DC/AC inverters, multilevel DC/AC inverters have many advantages such as lower switching flying voltage (from one level to next level), di/dt, and dv/dt, switching frequency, and THD. Unfortunately, most existing multilevel inverters contain too many components (independent and floating batteries and sources, diodes, capacitors, and switches). For example, a diode-clamped inverter, also called the neutral-point clamped (NPC) inverter, has $m = (2b + 1)$ levels and the components needed are

- $4b$ switches
- $2b$ capacitors
- $(4b - 2)$ diodes

where m is the level number, which is always an odd number, and b is the stage number (from the neutral point to the top point) [1–3]; a linear H-bridged inverter [2–4] has $m = (2b + 1)$ levels and the components needed are

- b floating batteries
- $4b$ switches
- $4b$ diodes

Figure 10.1 shows a five-level NPC inverter and a three-bridge inverter.

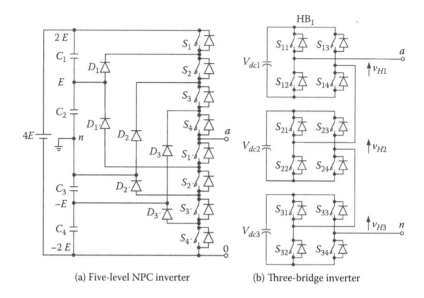

(a) Five-level NPC inverter   (b) Three-bridge inverter

FIGURE 10.1
Example inverters.

From Figure 10.1, we can see that a 5-level NPC inverter has a 4E battery (or 2×2 E batteries), 8 switches, 14 diodes, and 4 capacitors; a 7-level linear three-bridge inverter contains three floating batteries ($V_{dc1} = V_{dc2} = V_{dc3} = E$), 12 diodes, and 12 switches.

In this chapter, we will introduce laddered multilevel DC/AC inverters, which have a simple structure and straightforward operation and are a new development in the DC/AC inverter area. We will begin with a brief description on progressions.

10.2 Progressions (Series)

In mathematics, a progression is a series of numbers or quantities in which there is always the same relation between each quantity and the one succeeding it. We introduce several progressions in this section. In our laddered multilevel inverter design, we assume that in all progressions the value of the general ith term is V_i, and the value of the first term V_1 is 1.

10.2.1 Arithmetic Progressions

For all arithmetic progressions, the difference between the consecutive terms is constant. Some special cases are

TABLE 10.1

Unit Progression

Term no.	1	2	3	4	...	i	...	b
Value	1	1	1	1	...	1	...	1
S_i	1	2	3	4	...	i	...	b

- Unit progression, where all terms are equal to 1
- Natural progression, where all terms are natural numbers
- Odd number progression, where each term is odd-numbered

We define the value of the first term as V_1, the value of the general ith term as V_i, and the difference as d. Therefore, the value of the general term V_i is

$$V_i = V_1 + (i-1)d \tag{10.1}$$

The sum of the terms from the first term to the ith term is S_i,

$$S_i = iV_1 + \frac{i(i-1)}{2}d \tag{10.2}$$

10.2.1.1 Unit Progression

The unit progression is listed in Table 10.1, where $d = 0$. We assume that the last term is b with value V_b. From Equation (10.2), the sum of the terms S_i from 1 to the ith term is i.

10.2.1.2 Natural Number Progression

The natural number progression is listed in Table 10.2, where the difference $d = 1$ and the sum of the terms S_i from 1 to the ith term is $S_i = i + \frac{i(i-1)}{2}$.

10.2.1.3 Odd Number Progression

The odd number progression is listed in Table 10.3, where difference $d = 2$ and the sum of the terms S_i from 1 to the ith term is $S_i = i^2$.

TABLE 10.2

Natural Number Progression

Term no.	1	2	3	4	...	i	...	b
Value	1	2	3	4	...	i	...	b
S_i	1	3	6	10	...	$i + \frac{i(i-1)}{2}$	...	$b + \frac{b(b-1)}{2}$

TABLE 10.3

Odd Number Progression

Term no.	1	2	3	4	...	i	...	b
Value	1	3	5	7	...	$2i - 1$	...	$2b - 1$
S_i	1	4	9	16	...	i^2	...	b^2

10.2.2 Geometric Progressions

All geometric progressions have a constant ratio between successive terms. Two special series are

- Binary progression, where the ratio between successive terms equals 2
- Trinary progression, where the ratio equals 3

We define the ratio as r. Therefore, the value of the ith term V_i is

$$V_i = V_1 r^{i-1} \tag{10.3}$$

The sum of the terms S_i is

$$S_i = \frac{r^i - 1}{r - 1} V_1 \tag{10.4}$$

10.2.2.1 Binary Progression

The binary progression is listed in Table 10.4. Since the ratio $r = 2$, the sum of the terms S_i from 1 to the ith term is $S_i = 2^i - 1$.

10.2.2.2 Trinary Number Progression

The trinary progression is listed in Table 10.5. Since the ratio r is 3, the sum of the terms S_i from 1 to the ith term is $S_i = \frac{3^i - 1}{2}$.

10.2.3 New Progressions

Two new progressions are designed for laddered multilevel inverters. They are

- Luo progression
- Ye progression

TABLE 10.4

Binary Progression

Term no.	1	2	3	4	...	i	...	b
Value, V_i	1	2	4	8	...	2^{i-1}	...	2^{b-1}
Sum, S_i	1	3	7	15	...	$2^i - 1$	...	$2^b - 1$

TABLE 10.5

Trinary Progression

Term no.	1	2	3	4	...	i	...	b
Value, V_i	1	3	9	27	...	i	...	b
Sum, S_i	1	4	13	40	...	$\dfrac{3^i - 1}{2}$	...	$\dfrac{3^b - 1}{2}$

10.2.3.1 Luo Progression

The Luo progression is a new series that is different from any existing methods such as arithmetic and geometric progressions. The value of each term starting from the third term is twice the sum of the all previous terms plus 1:

$$V_i = \begin{cases} i & i \leq 2 \\ 7 \times 3^{i-2} & i \geq 3 \end{cases} \tag{10.5}$$

The sum of the terms from the first term to the ith term is S_i:

$$S_i = \sum_{k=1}^{i} V_k = \frac{7 \times 3^{i-2} - 1}{2} \quad i \geq 2 \tag{10.6}$$

The Luo progression is listed in Table 10.6. The sum of the terms from 1 to the ith term S_i is $S_i = \frac{7 \times 3^{i-2} - 1}{2}$ with $i \geq 2$.

10.2.3.2 Ye Progression

The Ye progression is another novel progression (series). The value of each term starting from the fourth is twice the sum of the all previous terms plus 1:

$$V_i = \begin{cases} i^2 - u(i - 2) & i \leq 3 \\ 25 \times 3^{i-4} & i \geq 4 \end{cases} \tag{10.7}$$

TABLE 10.6

Luo Progression

Term no.	1	2	3	4	...	$i \ (i \geq 3)$	...	b
Value, V_i	1	2	7	21	...	$7 \times 3^{i-3}$	...	$7 \times 3^{b-3}$
Sum, S_i	1	3	10	31	...	$\dfrac{7 \times 3^{i-2} - 1}{2}$	...	$\dfrac{7 \times 3^{b-2} - 1}{2}$

TABLE 10.7

Ye Progression

Term no.	1	2	3	4	5	...	i ($i≥4$)	...	b
Value, V_i	1	3	8	25	75	...	$25 \times 3^{i-4}$	...	$25 \times 3^{b-4}$
Sum, S_i	1	3	12	37	112	...	$\dfrac{25 \times 3^{i-3} - 1}{2}$	...	$\dfrac{25 \times 3^{b-3} - 1}{2}$

where $u(i - 2)$ is the unit-step function:

$$u(i-2) = \begin{cases} 0 & i = 1 \\ 1 & i \geq 2 \end{cases} \tag{10.8}$$

The sum of the terms from the first term to the ith term is S_i,

$$S_i = \sum_{k=1}^{i} V_k = \frac{25 \times 3^{i-3} - 1}{2} \quad i \geq 3 \tag{10.9}$$

The Ye progression is listed in Table 10.7. The sum of the terms from 1 to the ith term S_i is $S_i = \frac{25 \times 3^{i-3} - 1}{2}$ *with* $i \geq 3$.

10.3 Laddered Multilevel DC/AC Inverters

Before introducing laddered multilevel DC/AC inverters in this section, we start with some special switches.

10.3.1 Special Switches

10.3.1.1 Toggle Switch

The toggle switch is shown in Figure 10.2a, which is also called the one-pole two-throw (1P2T) switch. The switch has one wiper pole (W) and two contact

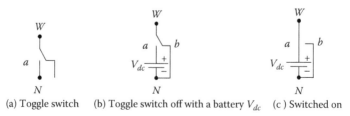

(a) Toggle switch   (b) Toggle switch off with a battery V_{dc} (c) Switched on

FIGURE 10.2
The toggle switch with a battery V_{dc}.

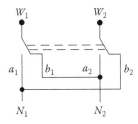

FIGURE 10.3
A two-pole two-throw (2P2T) switch.

positions "a" and "b." When the switch is on the wiper pole, it is linked to position "a," otherwise to position "b."

The terminal voltage of V_{WN} equals V_{dc} during switch-on as shown in Figure 10.2c, and 0 during switch-off as shown in Figure 10.2b.

10.3.1.2 Change-over Switch

A change-over switch is shown in Figure 10.3, which is also called the two-pole two-throw (2P2T) switch. It reverses the output voltage from the input voltage. If the input voltage is V_{W1W2} and the output voltage is V_{N1N2}, we have

$$V_{out} = V_{N1N2} = \begin{cases} V_{W1W2} & \text{switch is on} \\ -V_{W1W2} & \text{switch is off} \end{cases} \tag{10.10}$$

10.3.1.3 Band Switch

A band switch is shown in Figure 10.4, which is also called the one-pole multi-throw (1PMT) switch. It has one wiper pole (W) and multiple contact positions "0," "1," "2," ..., "m."

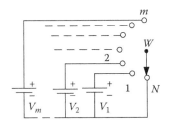

FIGURE 10.4
The band switch with M throws.

We assume that there are m voltage sources linked to the bands as V_1, V_2, ..., V_m, as shown in Figure 10.4. The terminal voltage of V_{WN} equals various source voltages:

$$V_{WN} = \begin{cases} 0 & \text{the wiper is on position } N \\ V_1 & \text{the wiper is on position } 1 \\ V_2 & \text{the wiper is on position } 2 \\ ... & ... \\ V_m & \text{the wiper is on position } m \end{cases} \qquad (10.11)$$

10.3.2 General Circuit of Laddered Inverters

The general circuit of laddered inverters is shown in Figure 10.5. It is symmetrical about the neutral point N; that is, there are two wings, positive and negative. Each wing has b sets of the toggle-switches S_i and batteries V_{dci}, where $i = -b, -(b-1), ..., -2, -1, 1, 2, ..., b-1$; b is the ladder stage number; m is the total level number, and i is the ith set number. The positive wing contains b sets in all, and the negative wing has the same number of sets by symmetry. Therefore, we have

$$|V_{dc-i}| = V_{dci} \quad i = 1, 2, ... b-1, b \qquad (10.12)$$

To simplify the analysis, we assume a purely resistive load R.

10.3.3 Linear Laddered Inverters (LLIs)

If all DC voltages V_{dci} in the laddered inverter shown in Figure 10.5 are the same voltage E, the result is a linear laddered inverter (LLI). This ladder was constructed following unit progression. The output voltage V_{out} has m levels and $m = 2b + 1$.

$$V_{dci} = E \quad i = -b, -(b-1), ... -2, -1, 1, 2, ... b-1, b \qquad (10.13)$$

The operation status:

$V_{out} = bE$: All positive wing switches on (others are off)
$V_{out} = (b-1)E$: Switches $S_1 - S_{b-1}$ are on (others are off)

......

$V_{out} = 2E$: Switches $S_1 - S_2$ are on (others are off)
$V_{out} = E$: only Switch S_1 is on (others are off)
$V_{out} = 0$: All Switches are off
$V_{out} = -E$: only Switch S_{-1} is on (others are off)

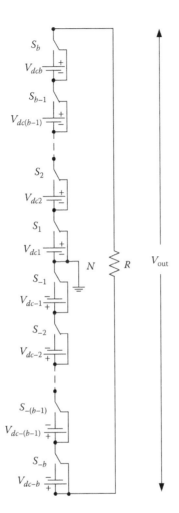

FIGURE 10.5
The general circuit of laddered inverters.

$V_{out} = -2E$: Switches $S_{-1} - S_{-2}$ are on (others are off)

......

$V_{out} = -(b-1)E$: Switches $S_{-1} - S_{-(b-1)}$ are on (others are off)
$V_{out} = -bE$: All negative wing switches on (others are off)

We obtain $m = 2b + 1$ levels. For example, if $b = 3$, we have the total level number $m = 7$. The output voltage waveform is shown in Figure 10.6.

10.3.4 Natural Number Laddered Inverters (NNLIs)

If all DC voltages V_{dci} in Figure 10.5 are iE, the inverter becomes a natural number laddered inverter (NNLI). This ladder was constructed following

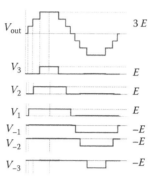

FIGURE 10.6
A seven-level output voltage waveform.

the natural number progression. The output voltage V_{out} has m levels, where $m = b^2 + b + 1$ (refer to Table 10.2):

$$V_{dci} = |i|E \quad i = -b, \ -(b-1), \ \dots -2, \ -1, \ 1, \ 2, \ \dots b-1, b \qquad (10.14)$$

The operation status:

- $V_{out} = \frac{b(b+1)}{2}E$: All positive wing switches on (the others are off)
- $V_{out} = \frac{b(b-1)}{2}E$: Switches S_2–S_b are on (the others are off)
- ......
- $V_{out} = 2E$: only Switch S_2 is on (others are off)
- $V_{out} = E$: only Switch S_1 is on (others are off)
- $V_{out} = 0$: All Switches are off
- $V_{out} = -E$: only Switch S_{-1} is on (others are off)
- $V_{out} = -2E$: only Switch S_{-2} is on (others are off)
- ......
- $V_{out} = -\frac{b(b-1)}{2}E$: Switches $S_{-2} - S_{-b}$ are on (others are off)
- $V_{out} = -\frac{b(b+1)}{2}E$: All negative wing switches on (others are off)

We obtain $m = b^2 + b + 1$ levels. For example, if $b = 3$, we have the total level number $m = 13$.

10.3.5 Odd Number Laddered Inverters (ONLIs)

If we set all DC voltages V_{dci} in the positive wing to be $(2i - 1)$ E, as shown in Figure 10.5, we have the odd number laddered inverter (ONLI). This ladder was constructed following an odd number progression. The output voltage V_{out} has m levels, and $m = 2b^2 + 1$ from Table 10.3:

$$V_{dci} = \begin{cases} (2i-1)E & i \geq 1 \\ (2i+1)E & i \leq -1 \end{cases} \tag{10.15}$$

The operation status:

- $V_{out} = b^2E$: All positive wing switches on (others are off)
- $V_{out} = (b^2 - 1)E$: Switches $S_2 - S_b$ are on (others are off)
- ......
- $V_{out} = 3E$: only Switch S_2 is on (others are off)
- $V_{out} = 2E$: Switch S_2 and S_{-1} are on (others are off)
- $V_{out} = E$: only Switch S_1 is on (others are off)
- $V_{out} = 0$: All Switches are off
- $V_{out} = -E$: only Switch S_{-1} is on (others are off)
- $V_{out} = -2E$: Switch S_{-2} and S_1 are on (others are off)
- $V_{out} = -3E$: only Switch S_{-2} is on (others are off)
- ......
- $V_{out} = -(b^2 - 1)E$: Switches $S_{-2} - S_{-b}$ are on (others are off)
- $V_{out} = -b^2E$: All negative wing switches on (others are off)

We obtain $m = 2b^2 + 1$ levels. For example, if $b = 3$, the total level number $m = 19$.

10.3.6 Binary Laddered Inverters (BLIs)

If we set all DC voltages V_{dci} in Figure 10.5 to be the voltage $(2^i - 1) E$, we obtain the binary laddered inverter (BLI). This ladder was constructed following a binary progression. The output voltage V_{out} has m levels, where $m = 2^{b+1} - 1$ from Table 10.4

The above DC voltages V_{dci} are binary voltages:

$$V_{dci} = \begin{cases} 2^{i-1}E & i \geq 1 \\ 2^{-i-1}E & i \leq -1 \end{cases} \tag{10.16}$$

The operation status:

- $V_{out} = (2^b - 1)E$: All positive wing switches on (others are off)
- ......
- $V_{out} = 3E$: Switches $S_1 - S_2$ are on (others are off)

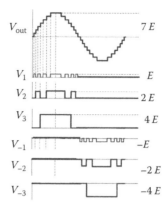

FIGURE 10.7
A fifteen-level output voltage waveform.

- $V_{out} = 2E$: only Switch S_2 is on (others are off)
- $V_{out} = E$: only Switch S_1 is on (others are off)
- $V_{out} = 0$: All Switches are off
- $V_{out} = -E$: only Switch S_{-1} is on (others are off)
- $V_{out} = -2E$: only Switch S_{-2} is on (others are off)
- $V_{out} = -3E$: Switches $S_{-1} - S_{-2}$ are on (others are off)
- ......
- $V_{out} = - (2^b - 1)E$: All negative wing switches on (others are off)

We achieved $m = 2^{b+1} - 1$ levels. For example, if $b = 3$, we have an $m = 15$ levels output voltage waveform as shown in Figure 10.7.

10.3.7 Modified Binary Laddered Inverters (MBLIs)

The modified binary laddered inverter (MBLI) is shown in Figure 10.8. We used a change-over (two-pole two-throw) switch to exchange the output voltage polarity, and saved half of the batteries and switches used in the negative wing.

For the operation status and output voltage waveform, see Section 10.3.6 and Figure 10.7. The output voltage V_{out} has m levels, where $m = 2^{b+1} - 1$.

10.3.8 Luo Progression Laddered Inverters (LPLIs)

For the Luo progression defined in Section 10.3.2.1,

$$V_i = \begin{cases} i & i \le 2 \\ 7 \times 3^{i-2} & i \ge 3 \end{cases} \tag{10.17}$$

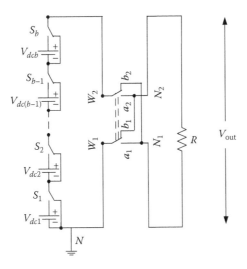

FIGURE 10.8
The modified binary laddered inverters.

the total level m, as shown in Table 10.8, is

$$m = \left(2 \times \sum_{i=1}^{b} V_i \right) + 1 = 7 \times 3^{b-2} \quad b \geq 2 \tag{10.18}$$

If we construct a ladder with 4 stages ($b = 4$) in both wings, from Table 10.8 we can obtain 63 levels ($m = 63$) in all. If the level unit is "E," the operation status is as shown below:

- $V_{out} = 31E$: All positive wing switches on (others are off)
- ……
- $V_{out} = 4E$: Switches S_3, S_{-1}, S_{-2} are on (others are off)
- $V_{out} = 3E$: Switches S_1 and S_2 are on (others are off)
- $V_{out} = 2E$: only Switch S_2 is on (others are off)
- $V_{out} = E$: only Switch S_1 is on (others are off)
- $V_{out} = 0$: All Switches are off

TABLE 10.8

Luo Progression

Stage no.	1	2	3	4	5	6	…	ith ($i \geq 3$)	…	b
Value, V_i	1	2	7	21	63	189	…	$7 \times 3^{i-3}$	…	$7 \times 3^{b-3}$
Sum, S_i	1	3	10	31	94	283	…	$\dfrac{7 \times 3^{i-2} - 1}{2}$	…	$\dfrac{7 \times 3^{b-2} - 1}{2}$
Total levels, m	3	7	21	63	189	567	…	$7 \times 3^{i-2}$	…	$7 \times 3^{b-2}$

- $V_{out} = -E$: only Switch S_{-1} is on (others are off)
- $V_{out} = -2E$: only Switch S_{-2} is on (others are off)
- $V_{out} = -3E$: Switches S_{-1} and S_{-2} are on (others are off)
- $V_{out} = -4E$: Switches, S_{-3}, S_1 and S_2 are on (others are off)
- ......
- $V_{out} = -31E$: All negative wing switches on (others are off)

10.3.9 Ye Progression Laddered Inverters (YPLIs)

For the Ye-progression defined in Section 10.3.2.2,

$$V_i = \begin{cases} i^2 - u(i-2) & i \leq 3 \\ 25 \times 3^{i-4} & i \geq 4 \end{cases} \tag{10.19}$$

the total level m, as shown in Table 10.9, is

$$n = \left(2 \times \sum_{i=1}^{b} V_i \right) + 1 = 25 \times 3^{b-4} \quad b \geq 4 \tag{10.20}$$

From Table 10.9, if we construct a ladder with 4 stages ($b = 4$) in both wings, we can obtain 75 levels ($m = 75$). If the level unit is "E," the operation status is as shown below:

- $V_{out} = 37E$: All positive wing switches on (others are off)
- ......
- $V_{out} = 4E$: Switches S_1 and S_3 are on (others are off)
- $V_{out} = 3E$: only Switch S_2 is on (others are off)
- $V_{out} = 2E$: Switches S_2 and S_{-1} are on (others are off)
- $V_{out} = E$: only Switch S_1 is on (others are off)
- $V_{out} = 0$: All Switches are off
- $V_{out} = -E$: only Switch S_{-1} is on (others are off)
- $V_{out} = -2E$: Switches S_{-2} and S_1 are on (others are off)

TABLE 10.9

Ye Progression

Stage no.	1	2	3	4	5	6	...	ith ($i \geq 4$)	...	b
Value, V_i	1	3	8	25	75	225	...	$25 \times 3^{i-4}$	...	$25 \times 3^{b-4}$
Sum, S_i	1	4	12	37	112	337	...	$\dfrac{25 \times 3^{i-3} - 1}{2}$	...	$\dfrac{25 \times 3^{b-3} - 1}{2}$
Total levels, n	3	9	25	75	225	675	...	$25 \times 3^{i-3}$	...	$25 \times 3^{b-3}$

- $V_{out} = -3E$: only Switch S_{-2} is on (others are off)
- $V_{out} = -4E$: Switches S_{-1} and S_{-2} are on (others are off)
- ......
- $V_{out} = -37E$: All negative wing switches on (others are off)

10.3.10 Trinary Laddered Inverters (TLIs)

If we set the DC voltages V_{dci} in Figure 10.5 to be trinary,

$$V_{dci} = \begin{cases} 3^{i-1}E & i \geq 1 \\ -3^{-i-1}E & i \leq 1 \end{cases} \tag{10.21}$$

the inverter becomes a trinary laddered inverter (TLI). The total level number $m = 3^b$. For example, if $b = 4$, we have the total level number $m = 81$.

The operation status is shown below:

- $V_{out} = 40E$: All positive wing switches on (others are off);
- ......
- $V_{out} = 3E$: only Switch S_2 is on (others are off)
- $V_{out} = 2E$: Switches S_2 and S_{-1} are on (others are off)
- $V_{out} = E$: only Switch S_1 is on (others are off)
- $V_{out} = 0$: All Switches are off
- $V_{out} = -E$: only Switch S_{-1} is on (others are off)
- $V_{out} = -2E$: Switches S_1 and S_{-2} are on (others are off)
- $V_{out} = -3E$: only Switch S_{-2} is on (others are off)
- ......
- $V_{out} = -40E$: All negative wing switches on (others are off)

10.4 Comparison of All Laddered Inverters

We introduced eight types of laddered inverters in Section 10.3. Table 10.10 shows a comparison of them with two other inverters, where the abbreviations are as follows:

- LLI—Linear laddered inverter
- NNLI—Natural number laddered inverter
- ONLI—Odd number laddered inverter
- BLI—Binary laddered inverter

TABLE 10.10

Comparison of All Laddered Inverters and Other Inverters

Inverters	LLI	NNLI	ONLI	BLI	MBLI	LPLI	YPLI	TLI	NPCI	LHBI
Stage no.	2b	2b	2b	2b	b	2b	2b	2b	2b	b
Battery no.	2b	2b	2b	2b	b	2b	2b	2b	1/2	b
Switch no.	2b	2b	2b	2b	b+1	2b	2b	2b	4b	4b
Capacitor no.	0	0	0	0	0	0	0	0	2b	0
Diode no.	0	0	0	0	0	0	0	0	4b − 2	4b
m, Total levels	2b + 1	$b^2 + b$ + 1	2b² + 1	2^{b+1} − 1	2^{b+1} − 1	$7 \times 3^{b-2}$	$25 \times 3^{b-3}$	3^b	2b + 1	2b + 1

- MBLI—Modified binary laddered inverter
- LPLI—Luo progression laddered inverter
- YPLI—Ye progression laddered inverter
- TLI—Trinary laddered inverter
- NPCI—Neutral-Point clamped inverter
- LHBI—linear H-bridged inverter

For example, if $b = 3$, the level numbers for each inverter are shown in Table 10.11.

When $b = 5$, the level numbers for each inverter are listed in Table 10.12, showing the extraordinary increase in the level numbers achieved and fewer

TABLE 10.11

Comparison of Inverters (if $b = 3$)

Inverters	LLI	NNLI	ONLI	BLI	MBLI	LPLI	YPLI	TLI	NPCI	LHBI
Stage no.	6	6	6	6	3	6	6	6	6	3
Battery no.	6	6	6	6	3	6	6	6	1/2	3
Switch no.	6	6	6	6	4	6	6	6	12	12
Capacitor no.	0	0	0	0	0	0	0	0	6	0
Diode no.	0	0	0	0	0	0	0	0	10	12
m, Total levels	7	13	19	15	15	21	25	27	7	7

TABLE 10.12

Comparison of Inverters (if $b = 5$)

Inverters	LLI	NNLI	ONLI	BLI	MBLI	LPLI	YPLI	TLI	NPCI	LHBI
Stage no.	10	10	10	10	5	10	10	10	10	5
Battery no.	10	10	10	10	5	10	10	10	1/2	5
Switch no.	10	10	10	10	6	10	10	10	20	20
Capacitor no.	0	0	0	0	0	0	0	0	10	0
Diode no.	0	0	0	0	0	0	0	0	18	20
n, Total levels	11	31	51	63	63	189	225	243	11	11

number of components used compared to those in Table 10.11, especially for the trinary ladder inverter (TLI), Ye progression laddered inverter (LPLI), and Luo progression laddered inverter (LPLI).

10.5 Solar Panel Energy Systems

The sun is a star in the universe. The earth is a planet, and flies in an oval orbit surrounding the sun. The sun is located at a focus of the oval orbit as shown in Figure 10.9 [4–9]. The average distance between the sun and earth is about 150 Mkm (150,000,000 kilometers). The sun radiates 3.8×10^{20} MW of energy into space. Earth receives 174×10^9 MW from the sun.

The solar panel is constructed with a solar cell (or photovoltaic cell), the technology of which embraces broad multidisciplinary subject areas. It converts solar energy into electricity by the photovoltaic effect. Briefly, solar cells are divided into two groups: monocrystalline and multicrystalline silicon wafers. Figure 10.10 shows a monocrystalline silicon wafer solar cell.

The solar panel shown in Figure 10.11 is assembled with a few solar cells. The theoretical power curve for a solar panel system is shown in Figure 10.12. The current is nearly constant in the low-voltage region. When the input voltage reaches 16.2 V, the input current sharply reduces to zero. The curve is the output power with its maximum power point (MPP) at 132 W.

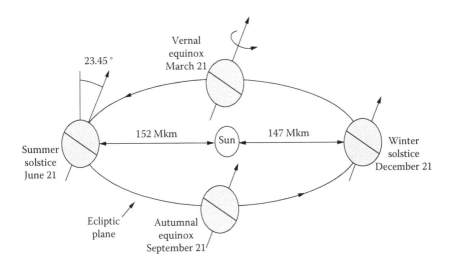

FIGURE 10.9
The sun and earth. (From Gilbert M. Masters 2004. *Renewable and Efficient Electric Power Systems.* New York: John Wiley & Sons. With permission.)

FIGURE 10.10
A monocrystalline solar cell.

FIGURE 10.11
A solar panel.

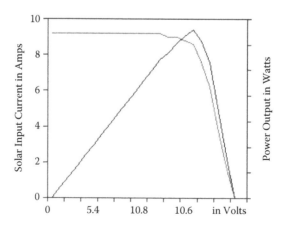

FIGURE 10.12
The theoretical system power curve.

We set the voltage unit $E = 16.2$ V. It is easy to construct the batteries for a BLI:

- $V_{dc1} = |V_{dc-1}| = E = 16.2$ V
- $V_{dc2} = |V_{dc-2}| = 2E = 32.4$ V
- $V_{dc3} = |V_{dc-3}| = 4E = 64.8$ V

Another setting to construct the batteries for an LPLI is the following:

- $V_{dc1} = |V_{dc-1}| = E = 16.2$ V
- $V_{dc2} = |V_{dc-2}| = 2E = 32.4$ V
- $V_{dc3} = |V_{dc-3}| = 7E = 113.4$ V

10.6 Simulation and Experimental Results

We use a BLI with $b = 3$ to do the simulation. The output voltage has 15 levels. The simulation and corresponding experimental results are shown in Figures 10.13 and 10.14, respectively.

We use an LPLI with $b = 3$ to do the simulation again. The output voltage has 21 levels. The simulation result is shown in Figure 10.15, and corresponding experimental result is shown in Figure 10.16.

Furthermore, we use the circuits for 41-level and 51-level inverters to do the simulation again. Their output voltages have 41 and 51 levels, respectively. Their simulation results are shown in Figures 10.17 and 10.19. Their corresponding experimental results are shown in Figures 10.18 and 10.20.

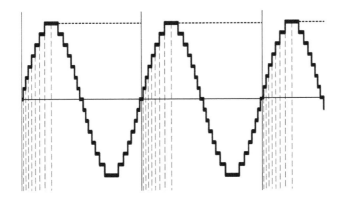

FIGURE 10.13
The simulation result of a fifteen-level BLI.

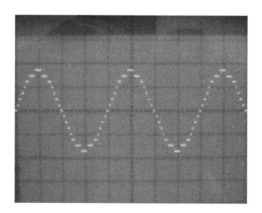

FIGURE 10.14
The experimental result of a 15-level BLI.

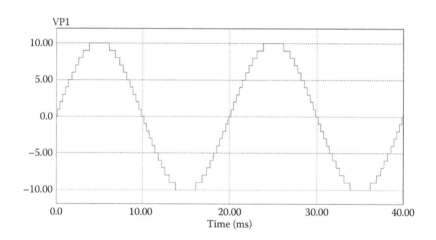

FIGURE 10.15
The simulation result of a 21-level LPLI.

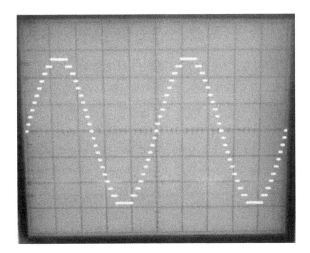

FIGURE 10.16
The experimental result of a 21-level LPLI.

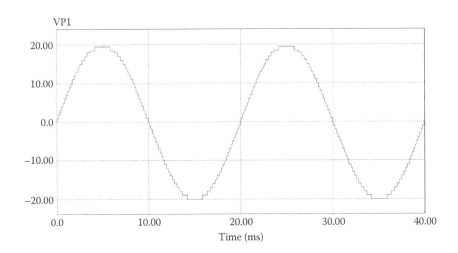

FIGURE 10.17
The simulation result of a 41-level inverter.

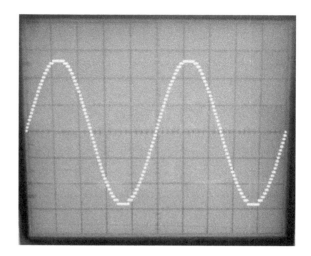

FIGURE 10.18
The experimental result of a 41-level inverter.

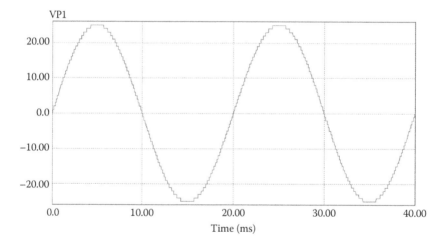

FIGURE 10.19
The simulation result of a 51-level inverter.

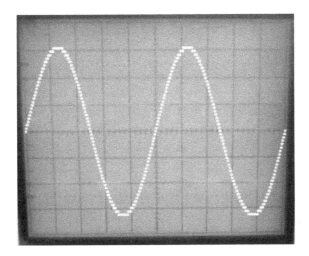

FIGURE 10.20
The experimental result of a fifty-one level inverter.

References

1. Luo F. L. 2011. Laddered multilevel DC/AC inverters used in solar panel energy systems. (invited keynote). *Proc. Int. Forum on the Energy Planning and Green Power Organization*, TaoYuan, Taiwan, pp. 1-25.
2. Luo F. L. and Ye H. 2010. *Power Electronics: Advanced Conversion Technologies*. Boca Raton, FL: Taylor & Francis.
3. Muhammad H. Rashid 2011. *Power Electronics Handbook (3rd edition)*. Boston: Butterworth-Heinemann.
4. Gilbert M. Masters 2004. *Renewable and Efficient Electric Power Systems*. New York: John Wiley & Sons.
5. Rashid, M. H. 2003. *Power Electronics: Circuits, Devices and Applications (3rd-edition)*. New Jersey: Prentice Hall.
6. Labouret A. and Villoz M. 2010. *Solar Photovoltaic Energy, the Institution of Engineering and Technology*. New Jersey: Prentice Hall.
7. Kalogirou S. A. 2009. *Solar Energy Engineering Processes and Systems*. San Diego: Academic Press.
8. Tagare D. M. 2010. *Electricity Power Generation*. New York: IEEE Press.
9. Kissell T. E. 2010. *Introduction to Solar Principles*. New Jersey, USA Prentice Hall.

11

Super-Lift Converter Multilevel DC/AC Inverters Used in Solar Panel Energy Systems

Super-lift converter multilevel DC/AC inverters will be introduced in this chapter. For convenience, we call them super-lift inverters or SL inverters (SLIs).

11.1 Introduction

The super-lift technique is the most outstanding contribution in DC/DC conversion technology [1–4]. Positive output super-lift Luo converters can easily lift the DC input voltage into a higher-level output DC voltage. In addition, positive output super-lift Luo converters have many subseries and circuits. The voltage transfer gains of the circuits increase in geometrical series (power series) stage by stage. Therefore, the super-lift technique has attracted much worldwide attention in recent decades. We were the first to apply super-lift Luo converters in DC/AC inverters.

An elementary positive output super-lift Luo converter is shown in Figure 11.1a. The equivalent circuits are shown in Figure 11.1b when switch S is on and Figure 11.1c when switch S is off.

When the main switch S is on, the inductor L_1 and capacitor C_1 are charged by the source voltage V_{in}. In the steady state, $V_{C1} = V_{in}$. When the main switch S is off, the output voltage V_O is highly lifted from source voltage by inductor L_1 and capacitor C_1, as can be seen from the following equation

$$V_O = \frac{2-k}{1-k} V_{in} \tag{11.1}$$

where k is the duty cycle. A re-lift positive output super-lift Luo converter is shown in Figure 11.2.

Its output voltage V_O is

$$V_O = \left(\frac{2-k}{1-k}\right)^2 V_{in} \tag{11.2}$$

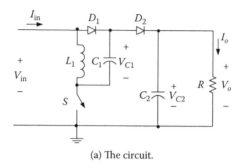

(a) The circuit.

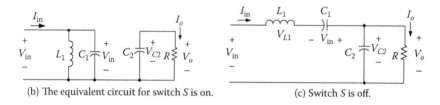

(b) The equivalent circuit for switch S is on.   (c) Switch S is off.

FIGURE 11.1
Elementary circuit of a positive output super-lift Luo converter.

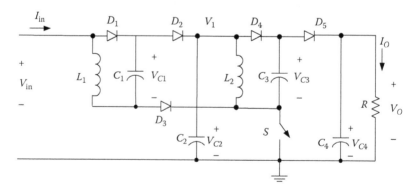

FIGURE 11.2
Re-lift circuit of the positive output super-lift Luo converter.

11.2 Super-Lift Converter Used in Multilevel DC/AC Inverters

In this section, we show how to apply the super-lift conversion technique in multilevel DC/AC inverters.

11.2.1 Seven-Level SL Inverter

A seven-level super-lift inverter is shown in Figure 11.3.

There is only one DC voltage source E needed in this circuit. A change-over-switch (2P2T) and a three-position band-switch are used in the circuit. When $k = 0.5$, the output voltage is 3E. Three capacitors, C_2, C_3, and C_4, are used to split the output voltage in 3 levels: E, 2E, and 3E. Therefore, the operation status is as follows:

- $V_{out} = 3E$: The change-over-switch is on, and the band-switch is at position 3.
- $V_{out} = 2E$: The change-over-switch is on, and the band-switch is at position 2.
- $V_{out} = E$: The change-over-switch is on, and the band-switch is at position 1.
- $V_{out} = 0$: The band-switch is at position 0 (i.e., N).
- $V_{out} = -E$: The change-over-switch is off, and the band-switch is at position 1.
- $V_{out} = -2E$: The change-over-switch is off, and the band-switch is at position 2.
- $V_{out} = -3E$: The change-over-switch is off, and the band-switch is at position 3.

We have obtained a seven-level output AC voltage. The output voltage peak value is three times the input DC voltage E. The waveform is shown in Figure 11.4.

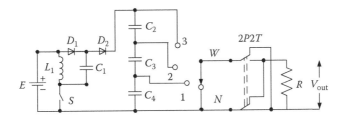

FIGURE 11.3
A seven-level super-lift inverter.

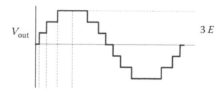

FIGURE 11.4
Seven-level waveform.

11.2.2 Fifteen-Level SL Inverter

A 15-level super-lift inverter is shown in Figure 11.5.

One change-over-switch (2P2T) is used in the circuit. When $k = 6/7$ (≈ 0.857), the output voltage is 7E. We use seven capacitors, C_2, C_3, C_4, C_5, C_6, C_7, and C_8, to split the output voltage in seven levels: E, 2E, 3E, 4E, 5E, 6E, and 7E. Therefore, the operation status is as follows:

- $V_{out} = 7E$: The change-over-switch is on, and the band-switch is at position 7.

- $V_{out} = 6E$: The change-over-switch is on, and the band-switch is at position 6.

- $V_{out} = 5E$: The change-over-switch is on, and the band-switch is at position 5.

- $V_{out} = 4E$: The change-over-switch is on, and the band-switch is at position 4.

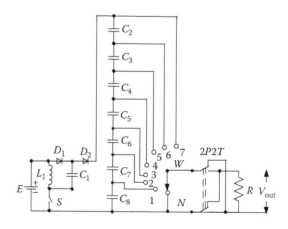

FIGURE 11.5
A fifteen-level super-lift inverter.

- $V_{out} = 3E$: The change-over-switch is on, and the band-switch is at position 3.
- $V_{out} = 2E$: The change-over-switch is on, and the band-switch is at position 2.
- $V_{out} = E$: The change-over-switch is on, and the band-switch is at position 1.
- $V_{out} = 0$: The band-switch is at position 0 (i.e., N).
- $V_{out} = -E$: The change-over-switch is off, and the band-switch is at position 1.
- $V_{out} = -2E$: The change-over-switch is off, and the band-switch is at position 2.
- $V_{out} = -3E$: The change-over-switch is off, and the band-switch is at position 3.
- $V_{out} = -4E$: The change-over-switch is off, and the band-switch is at position 4.
- $V_{out} = -5E$: The change-over-switch is off, and the band-switch is at position 5.
- $V_{out} = -6E$: The change-over-switch is off, and the band-switch is at position 6.
- $V_{out} = -7E$: The change-over-switch is off, the band-switch is at position 7.

We have obtained a 15-level output AC voltage. The output voltage peak value is seven times the input DC voltage E. The waveform is shown in Figure 11.6.

11.2.3 Twenty-One-Level SC Inverter

A positive output re-lift super-lift Luo converter is used to construct a 21-level inverter. Its circuit diagram is shown in Figure 11.7. A change-over switch (2P2T) is used in the circuit.

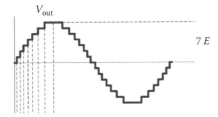

FIGURE 11.6
A fifteen-level waveform.

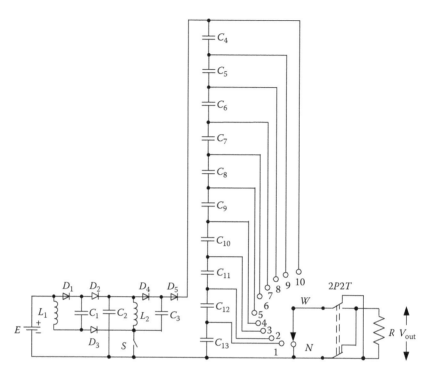

FIGURE 11.7
A 21-level super-lift inverter.

When $k = \frac{\sqrt{10}-2}{\sqrt{10}-1}$ ($k \approx 0.54$), the output voltage is 10E. We use 10 capacitors, C_4, C_5, C_6, C_7, C_8, C_9, C_{10}, C_{11}, C_{12}, and C_{13}, to split the output voltage in 10 levels: E, 2E, 3E, 4E, 5E, 6E, 7E, 8E, 9E, and 10E. Therefore, the operation status is as follows:

- $V_{out} = 10E$: The change-over-switch is on, and the band-switch is at position 10.
- $V_{out} = 9E$: The change-over-switch is on, and the band-switch is at position 9.
- $V_{out} = 8E$: The change-over-switch is on, and the band-switch is at position 8.
- $V_{out} = 7E$: The change-over-switch is on, and the band-switch is at position 7.
- $V_{out} = 6E$: The change-over-switch is on, and the band-switch is at position 6.
- $V_{out} = 5E$: The change-over-switch is on, and the band-switch is at position 5.
- $V_{out} = 4E$: The change-over-switch is on, and the band-switch is at position 4.

- $V_{out} = 3E$: The change-over-switch is on, and the band-switch is at position 3.
- $V_{out} = 2E$: The change-over-switch is on, and the band-switch is at position 2.
- $V_{out} = E$: The change-over-switch is on, and the band-switch is at position 1.
- $V_{out} = 0$: The band-switch is at position 0 (i.e., N).
- $V_{out} = -E$: The change-over-switch is off, and the band-switch is at position 1.
- $V_{out} = -2E$: The change-over-switch is off, and the band-switch is at position 2.
- $V_{out} = -3E$: The change-over-switch is off, and the band-switch is at position 3.
- $V_{out} = -4E$: The change-over-switch is off, and the band-switch is at position 4.
- $V_{out} = -5E$: The change-over-switch is off, and the band-switch is at position 5.
- $V_{out} = -6E$: The change-over-switch is off, and the band-switch is at position 6.
- $V_{out} = -7E$: The change-over-switch is off, and the band-switch is at position 7.
- $V_{out} = -8E$: The change-over-switch is off, and the band-switch is at position 8.
- $V_{out} = -9E$: The change-over-switch is off, and the band-switch is at position 9.
- $V_{out} = -10E$: The change-over-switch is off, and the band-switch is at position 10.

We have obtained a 21-level output AC voltage. The output voltage peak value is 10 times the input DC voltage E. The waveform is shown in Figure 11.8.

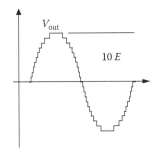

FIGURE 11.8
A 21-level waveform.

Similarly, higher-level inverters can be constructed by using higher-stage super-lift converters.

11.3 Simulation and Experimental Results

A 15-level super-lift inverter in a solar panel energy system is an example for simulation. The output voltage of a 15-level SL inverter is shown in Figure 11.9. The corresponding experimental result is shown in Figure 11.10.

For a 25-level super lift inverter in the solar panel energy system, the simulation and corresponding experimental results are shown in Figures 11.11 and 11.12, respectively.

Furthermore, we use the super-lift conversion technique to produce 35-level and 45-level super lift inverters for the solar panel energy system. Their output voltages have 35 and 45 levels, respectively. The simulation and experimental results are shown in Figures 11.13–11.16.

We introduced super-lift multilevel inverters in this chapter, which is a new approach to DC/AC inversion technology in the field. All super-lift multilevel inverters have a simple structure, straightforward operation procedure, easy control, and high output voltage. Fewer components are used to construct higher-level output voltage, and therefore the cost is lower. We have applied four SLIs from 15-level to 45-level of output voltage to a solar panel energy system and obtained satisfactory simulation and experimental results. These super-lift multilevel inverters can also be used in other renewable energy systems and industrial applications.

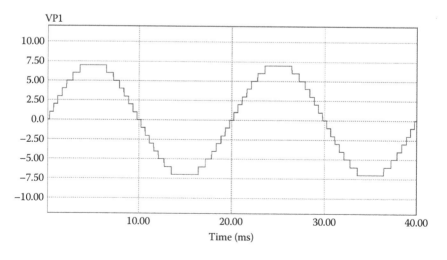

FIGURE 11.9
The simulation result of a fifteen-level SLI.

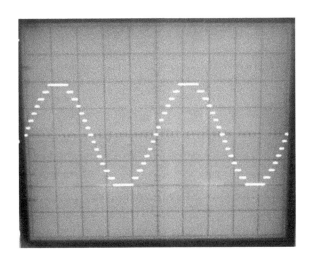

FIGURE 11.10
The experimental result of a fifteen-level SLI.

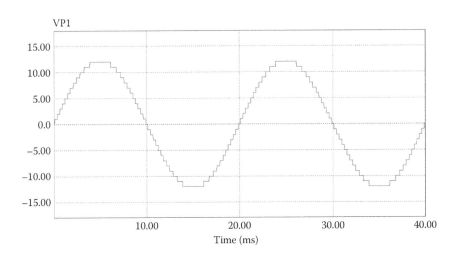

FIGURE 11.11
The simulation result of a 25-level SLI.

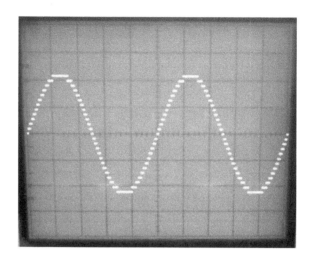

FIGURE 11.12
The experimental result of a 25-level SLI.

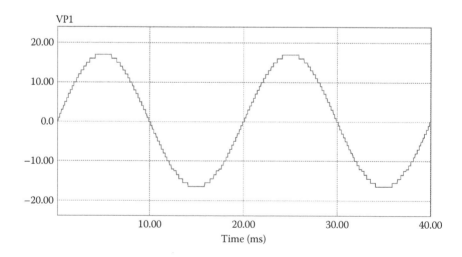

FIGURE 11.13
The simulation result of a thirty-five level SLI.

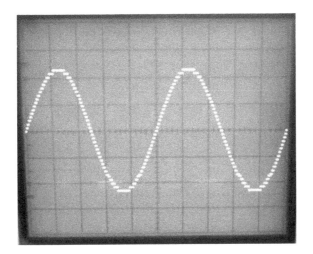

FIGURE 11.14
The experimental result of a 35-level SLI.

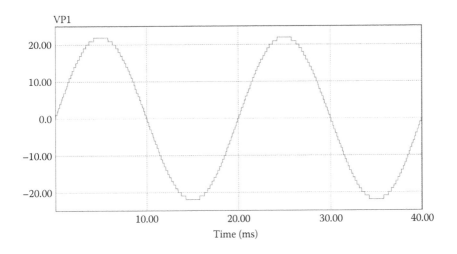

FIGURE 11.15
The simulation result of a 45-level SLI.

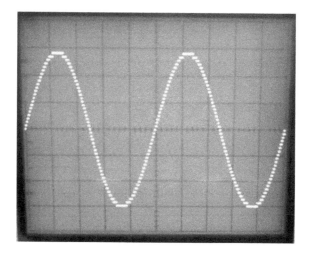

FIGURE 11.16
The experimental result of a 45-level SLI.

References

1. Luo F. L. and Ye H., 2004. *Advanced DC/DC Converters.* Boca Raton, FL: CRC Press.
2. Luo F. L. and Ye H. 2010. *Power Electronics: Advanced Conversion Technologies.* Boca Raton, FL: Taylor & Francis.
3. Luo F. L. and Ye H. 2002. Super-lift Luo converter. *Proc. IEEE Int. Conf. PESC'2002,* Cairns, Australia, 23–27 June, pp. 425–430.
4. Luo F. L. and Ye H. 2003. Positive output super-lift converters. *IEEE Trans. Power Electron.,* Vol. 18, No. 1, January, pp. 105–113.

12

Switched-Capacitor Multilevel DC/AC Inverters in Solar Panel Energy Systems

Switched-capacitor multilevel DC/AC inverters will be described in this chapter. For convenience, we call them switched-capacitor inverters or SC inverters (SCIs).

12.1 Introduction

A switched capacitor (SC) is usually manufactured with a switch and a capacitor together [1–6]. It has been used in DC/DC converters for many years. It can be integrated into power semiconductor IC chips. Hence, SC converters have small size and work at high switching frequencies. This technique opened the way to building converters with high power density and attracted the attention of research workers and manufacturers. We were the first to use switched capacitors in DC/AC inverters.

A switched-capacitor DC/DC converter is shown in Figure 12.1a. It contains two SCs (C_1 and C_2), main switch S, two slave switches (S_1 and S_2), and three diodes. The main switch S and the slave switches are operated mutually exclusively; that is, when the main switch is on, the slave switches are off, and vice versa.

When the main switch S is on, the slave switches are off, and all diodes conduct. The equivalent circuit is shown in Figure 12.1b. Both SCs are charged by the source voltage E in the steady state. When the main switch S is off, the slave switches are on, and all diodes are blocked. The equivalent circuit is shown in Figure 12.1c. The voltages at the points 1, 2, and 3 are E, 2E, and 3E, respectively, in the steady state.

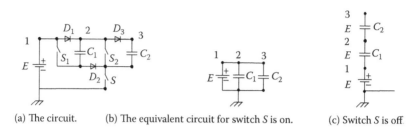

(a) The circuit.   (b) The equivalent circuit for switch S is on.   (c) Switch S is off

FIGURE 12.1
A switched capacitor.

12.2 Switched Capacitor Used in Multilevel DC/AC Inverters

In this section, we show how to apply the switched capacitor technique in multilevel DC/AC inverters.

12.2.1 Five-Level SC Inverter

A 5-level switched capacitor inverter is shown in Figure 12.2.

There are one DC voltage source E, one 3-position band-switch, and one change-over switch (2P2T) in the circuit. The slave switch S_1 and the main switch S switch mutually exclusively; that is, when S is on, S_1 is off, and vice versa. Capacitor C_1 is a switched capacitor. When S is on and S_1 is off, diode D_1 conducts. Therefore, Capacitor C_1 is charged to the voltage E in steady state. When S is off and S_1 is on, diode D_1 is blocked. The voltage at point 2 is $V_2 = 2 \times E$ (V_1 always equals E). Therefore, the operation status is as follows:

- $V_{out} = 2E$: 2P2T is on, the band switch is at position 2, and the main switch S is off.

- $V_{out} = E$: 2P2T is on, the band switch is at position 1, and the main switch S is on.

- $V_{out} = 0$: The band switch is at position 0 (i.e., N), and all switches can be on or off.

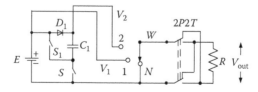

FIGURE 12.2
A five-level switched-capacitor inverter.

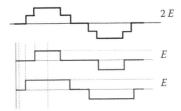

FIGURE 12.3
A five-level waveform.

- $V_{out} = -E$: 2P2T is off, the band switch is at position 1, and the main switch S is on.
- $V_{out} = -2E$: 2P2T is off, the band switch is at position 2, and the main switch S is off.

We have obtained a five-level output AC voltage. The output voltage peak value is two times the input DC voltage E. The waveform is shown in Figure 12.3.

12.2.2 Nine-Level SC Inverter

A nine-level switched-capacitor inverter is shown in Figure 12.4.

There is one DC voltage source E, one five-position band switch, and one change-over switch (2P2T) switch in the circuit. The slave switches S_{1-3} and the main switch S switch mutually exclusively; that is, when S is on, all slave switches are off, and vice versa. Capacitors C_{1-3} are the switched capacitors. When S is on, all diodes conduct. Therefore, all SCs are charged to the voltage E in the steady state. When S is off and S_1 is on, diode D_1 is blocked. The voltage at point 2 is $V_2 = 2 \times E$; the voltage at point 3 is $V_3 = 3 \times E$; the voltage at point 4 is $V_4 = 4 \times E$; (V_1 is always E). Therefore, the operation status is as follows:

- $V_{out} = 4E$: 2P2T is on, the band switch is at position 4, and the main switch S is off.
- $V_{out} = 3E$: 2P2T is on, the band switch is at position 3, and the main switch S is off.

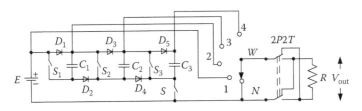

FIGURE 12.4
A nine-level switched-capacitor inverter.

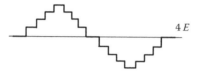

FIGURE 12.5
A nine-level waveform.

- $V_{out} = 2E$: 2P2T is on, the band switch is at position 2, and the main switch S is off.
- $V_{out} = E$: 2P2T is on, the band switch is at position 1, and the main switch S is on.
- $V_{out} = 0$: The band switch is at position 0 (i.e., N), and all switches can be on or off.
- $V_{out} = -E$: 2P2T is off, the band switch is at position 1, and the main switch S is on.
- $V_{out} = -2E$: 2P2T is off, the band switch is at position 2, and the main switch S is off.
- $V_{out} = -3E$: 2P2T is off, the band switch is at position 3, and the main switch S is off.
- $V_{out} = -4E$: 2P2T is off, the band switch is at position 4, and the main switch S is off.

We have obtained a nine-level output AC voltage. The output voltage peak value is four times the input DC voltage E. The waveform is shown in Figure 12.5.

12.2.3 Fifteen-Level SC Inverter

A 15-level switched-capacitor inverter is shown in Figure 12.6.

There is one DC voltage source E, one 7-position band switch, and one change-over switch (2P2T) switch in the circuit. The slave switches S_{1-6} and the main switch S switch mutually exclusively; that is, when S is on, all slave switches off, and vice versa. Capacitors C_{1-6} are SCs. When S is on and all slave switches are off, all diodes conduct. Therefore, all SCs are charged to

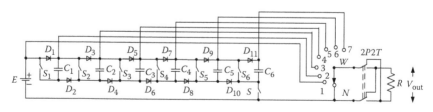

FIGURE 12.6
A fifteen-level switched-capacitor inverter.

the voltage E in the steady state. The voltage at point 2 is $V_2 = 2 \times E$; the voltage at point 2 is $V_3 = 3 \times E$; the voltage at point 4 is $V_4 = 4 \times E$, and so on, where V_1 is always E. Therefore, the operation status is as follows:

- V_{out} = 7E: 2P2T is on, the band switch is at position 7, and the main switch S is off.
- V_{out} = 6E: 2P2T is on, the band switch is at position 6, and the main switch S is off.
- V_{out} = 5E: 2P2T is on, the band switch is at position 5, and the main switch S is off.
- V_{out} = 4E: 2P2T is on, the band switch is at position 4, and the main switch S is off.
- V_{out} = 3E: 2P2T is on, the band switch is at position 3, and the main switch S is off.
- V_{out} = 2E: 2P2T is on, the band switch is at position 2, and the main switch S is off.
- V_{out} = E: 2P2T is on, the band switch is at position 1, and the main switch S is on.
- V_{out} = 0: The band switch is at position 0 (i.e., N), and all switches are on.
- V_{out} = –E: 2P2T is off, the band switch is at position 1, and the main switch S is on.
- V_{out} = –2E: 2P2T is off, the band switch is at position 2, and the main switch S is off.
- V_{out} = –3E: 2P2T is off, the band switch is at position 3, and the main switch S is off.
- V_{out} = –4E: 2P2T is off, the band switch is at position 4, and the main switch S is off.
- V_{out} = –5E: 2P2T is off, the band switch is at position 5, and the main switch S is off.
- V_{out} = –6E: 2P2T is off, the band switch is at position 6, and the main switch S is off.
- V_{out} = –7E: 2P2T is off, the band switch is at position 7, and the main switch S is off.

We have obtained a 15-level output AC voltage. The output voltage peak value is seven times the input DC voltage E. The waveform is shown in Figure 12.7.

12.2.4 Higher-Level SC Inverter

Repeatedly adding components (S_1-C_1-D_1-D_2) as shown in Figure 12.6, we can obtain higher-level inverters. We believe that readers of this book have

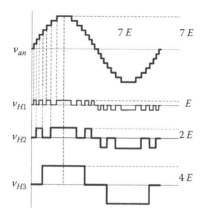

FIGURE 12.7
A fifteen-level waveform.

understood how to construct higher-level inverters, for example, a 21-level SC inverter.

12.3 Simulation and Experimental Results

Switched-capacitor multilevel inverters in solar panel energy systems are examples for the simulation. The 17-level inverter's simulation result is shown in Figure 12.8. Its corresponding experimental result is shown in Figure 12.9.

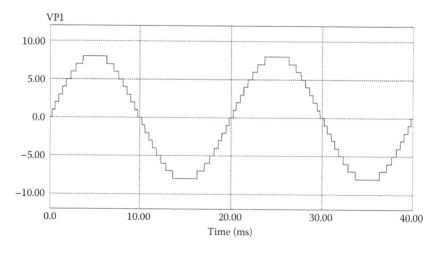

FIGURE 12.8
A seventeen-level simulation waveform.

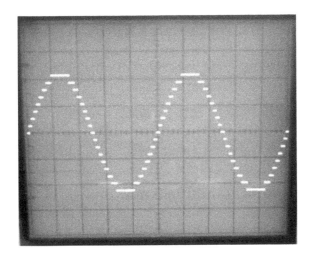

FIGURE 12.9
A seventeen-level experimental waveform.

The 27-level inverter's simulation and corresponding experimental results can be seen in Figures 12.10 and 12.11, respectively.

Furthermore, we use the switched-capacitor technique to produce 37-level and 47-level SC inverters for the solar panel energy system. Their output voltages have 37 and 47 levels, respectively. Their simulation and experimental results are shown in Figures 12.12–12.15.

We introduced switched-capacitor multilevel inverters in this chapter. All SC multilevel inverters have relatively simple structure, straightforward

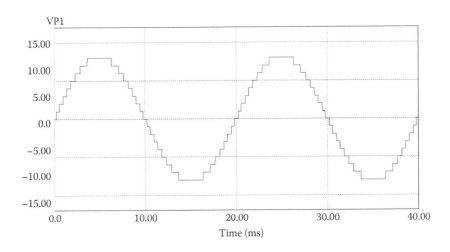

FIGURE 12.10
A 27-level simulation waveform.

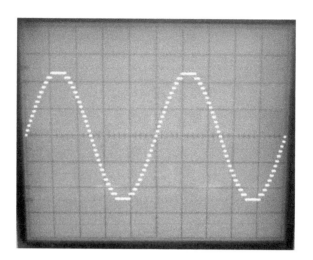

FIGURE 12.11
A 27-level experimental waveform.

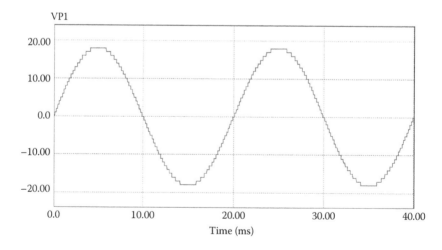

FIGURE 12.12
A 37-level simulation waveform.

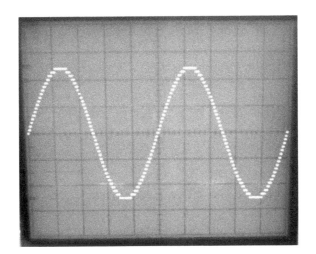

FIGURE 12.13
A 37-level experimental waveform.

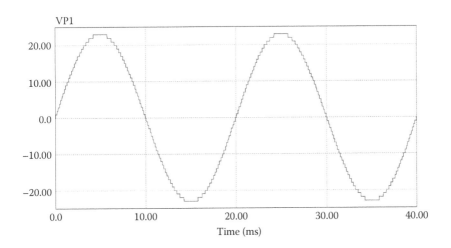

FIGURE 12.14
A 47-level simulation waveform.

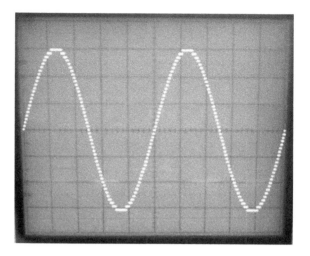

FIGURE 12.15
A 47-level experimental waveform.

operation procedure, easy control, and higher output voltage (compared with the input voltage). We can use fewer components to construct more levels of the output voltage. We applied four SCIs from 17-level to 47-level of output voltage to a solar panel energy system and obtained satisfactory simulation and experimental results that strongly supported our circuit design. These SC multilevel inverters can be used in other renewable energy systems and industrial applications.

References

1. Luo F. L. and Ye H. 2010. *Power Electronics: Advanced Conversion Technologies.* Boca Raton, FL: Taylor & Francis.
2. Luo F. L. and Ye H. 2004. *Advanced DC/DC Converters.* Boca Raton, FL: CRC Press.
3. Luo F. L. and Ye H. 2004. Positive output multiple-lift push-pull switched-capacitor Luo converters. *IEEE Trans. Ind. Electron.,* Vol. 51, No. 3, pp. 594–602.
4. Gao Y. and Luo F. L. 2001. Theoretical analysis on performance of a 5V/12V push-pull switched capacitor DC/DC converter. *Proc. IEE Int. Conf. IPEC'2001,* Singapore, 17–19 May, pp. 711–715.
5. Luo F. L., Ye H., and Rashid M. H. 1999. Four-quadrant switched capacitor Luo converter. *Int. J. Power Supply Technol. Applicat.,* Vol. 2, No. 3, June, pp. 4–10.
6. Luo F. L. and Ye H. 1999. Two-quadrant switched capacitor converter. *Proc. 13th Chinese Power Supply Society IAS Annual Meeting,* Shenzhen, China, pp. 164–168.

13

Switched Inductor Multilevel DC/AC Inverters Used in Solar Panel Energy Systems

Switched inductor multilevel DC/AC inverters will be described in this chapter. For convenience, we call them switched inductor inverters or SI inverters (SIIs).

13.1 Introduction

Although switched capacitor inverters can reach high power density, their circuits are relatively complex with many switches and control circuitries [1–3]. If the difference between input and output voltages is large, multiple switched capacitor stages must be employed. Switched inductor inverters successfully overcame this disadvantage. Usually, only one inductor is required for each inverter no matter how large the difference between the input and output voltages. Therefore, the switched inductor inverter has very simple circuit and, consequently, very high power density.

A switched inductor inverter is shown in Figure 13.1a. It contains one inverter L, main switch S, and one Diode D. Figures 13.1b and 13.1c show the equivalent circuit when switch S is on and off, respectively.

13.2 Switched Inductor Used in Multilevel DC/AC Inverters

In this section, we show how to apply the switched inductor technique in multilevel DC/AC inverters.

13.2.1 Five-Level SI Inverter

A five-level switched inductor inverter is shown in Figure 13.2.

There is one DC voltage source E, one three-position band switch, and one change-over switch (2P2T) in the circuit. The main switch S is on or off with

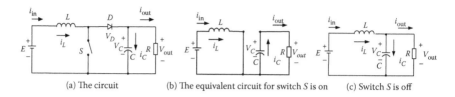

(a) The circuit   (b) The equivalent circuit for switch S is on   (c) Switch S is off

FIGURE 13.1
A switched inductor inverter.

the duty cycle $k = 0.5$ to charge the two capacitors C_1 and C_2 with voltage E for each. Therefore, the operation status is as follows:

- $V_{out} = 2E$: 2P2T is on, and the band switch is at position 2.
- $V_{out} = E$: 2P2T is on, and the band switch is at position 1.
- $V_{out} = 0$: The band switch is at position 0 (i.e., N).
- $V_{out} = -E$: 2P2T is off, and the band switch is at position 1.
- $V_{out} = -2E$: 2P2T is off, and the band switch is at position 2.

We have obtained a five-level output AC voltage. The output voltage peak value is two times the input DC voltage E. The waveform is shown in Figure 13.3.

13.2.2 Nine-Level SL Inverter

A nine-level switched-capacitor inverter is shown in Figure 13.4.

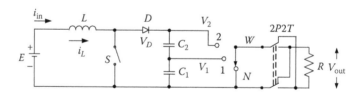

FIGURE 13.2
A five-level switched-inductor inverter.

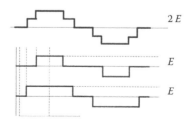

FIGURE 13.3
A five-level waveform.

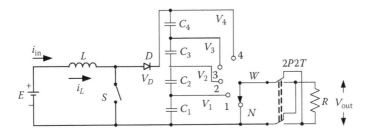

FIGURE 13.4
A nine-level switched-inductor inverter.

There is one DC voltage source E, one five-position band switch, and one change-over switch (2P2T) in the circuit. The main switch S is on/off with the duty cycle $k = 0.2$ to charge the four capacitors C_1, C_2, C_3, and C_4 with voltage E for each. Therefore, the operation status is as follows:

- $V_{out} = 4E$: 2P2T is on, and the band switch is at position 4.
- $V_{out} = 3E$: 2P2T is on, and the band switch is at position 3.
- $V_{out} = 2E$: 2P2T is on, and the band switch is at position 2.
- $V_{out} = E$: 2P2T is on, and the band switch is at position 1.
- $V_{out} = 0$: The band switch is at position 0 (i.e., N).
- $V_{out} = -E$: 2P2T is off, and the band switch is at position 1.
- $V_{out} = -2E$: 2P2T is off, and the band switch is at position 2.
- $V_{out} = -3E$: 2P2T is off, and the band switch is at position 3.
- $V_{out} = -4E$: 2P2T is off, and the band switch is at position 4.

We have obtained a nine-level output AC voltage. The output voltage peak value is four times the input DC voltage E. The waveform is shown in Figure 13.5.

13.2.3 Fifteen-Level SC Inverter

A 15-level switched-capacitor inverter is shown in Figure 13.6.

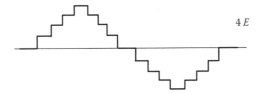

FIGURE 13.5
A nine-level waveform.

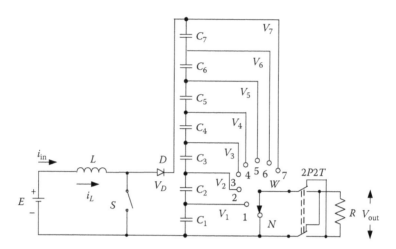

FIGURE 13.6
A fifteen-level switched-inductor inverter.

There is one DC voltage source E, one seven-position band switch, and one change-over switch (2P2T) in the circuit. The main switch S is on/off with the duty cycle $k = 6/7$ to charge the seven capacitors with voltage E for each. Therefore, all capacitors are charged to the voltage E in the steady state. The voltage at point 2 is $V_2 = 2 \times E$; at point 3 $V_3 = 3 \times E$; at point 4 $V_4 = 4 \times E$, and so on; (V_1 is E). Therefore, the operation status is as follows:

- $V_{out} = 7E$: 2P2T is on, and the band switch is at position 7.
- $V_{out} = 6E$: 2P2T is on, and the band switch is at position 6.
- $V_{out} = 5E$: 2P2T is on, and the band switch is at position 5.
- $V_{out} = 4E$: 2P2T is on, and the band switch is at position 4.
- $V_{out} = 3E$: 2P2T is on, and the band switch is at position 3.
- $V_{out} = 2E$: 2P2T is on, and the band switch is at position 2.
- $V_{out} = E$: 2P2T is on, and the band switch is at position 1.
- $V_{out} = 0$: The band-switch is at position 0 (i.e., N).
- $V_{out} = -E$: 2P2T is off, and the band switch is at position 1.
- $V_{out} = -2E$: 2P2T is off, and the band switch is at position 2.
- $V_{out} = -3E$: 2P2T is off, and the band switch is at position 3.
- $V_{out} = -4E$: 2P2T is off, and the band switch is at position 4.
- $V_{out} = -5E$: 2P2T is off, and the band switch is at position 5.
- $V_{out} = -6E$: 2P2T is off, and the band switch is at position 6.
- $V_{out} = -7E$: 2P2T is off, and the band switch is at position 7.

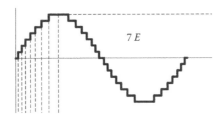

FIGURE 13.7
A fifteen-level waveform.

We have obtained a fifteen-level output AC voltage. The output voltage peak value is seven times the input DC voltage E. The waveform is shown in Figure 13.7.

Repeatedly adding the components in Figure 13.6, we can obtain higher-level inverters.

13.3 Simulation and Experimental Results

Switched inductor multilevel inverters in a solar panel energy system are examples for the simulation. The nine-level SI inverter has an output voltage with nine levels. The simulation result is shown in Figure 13.8. Its corre-

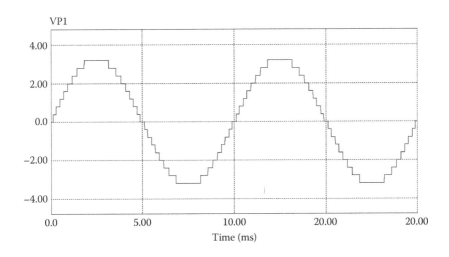

FIGURE 13.8
A nine-level simulation waveform.

sponding experimental result is shown in Figure 13.9. The 19-level SI inverter

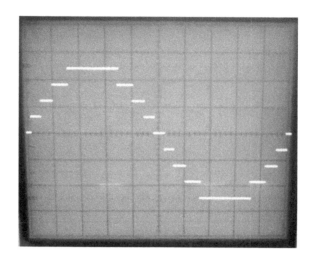

FIGURE 13.9
A nine-level experimental waveform.

has an output voltage with 19 levels. The simulation result is shown in Figure 13.10. The corresponding experimental result is shown in Figure 13.11

Furthermore, we use the switched-inductor technique to produce 39-level and 49-level SI inverters for the solar panel energy system. Their output voltages have 39 levels and 49 levels, respectively. Their simulation and corresponding experimental results are shown in Figures 13.12–13.15.

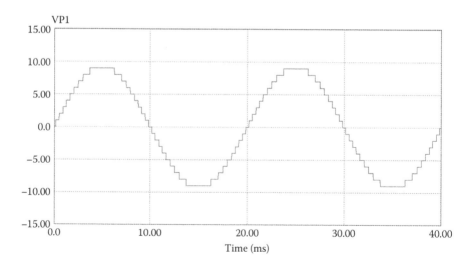

FIGURE 13.10
A 19-level simulation waveform.

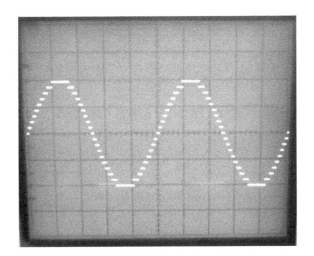

FIGURE 13.11
A 19-level experimental waveform.

We have introduced switched inductor multilevel inverters in this chapter. All SL multilevel inverters have advantages such as very simple structure, straightforward operation procedure, easy control, and higher output voltage (compared with the input voltage). Fewer components are needed to construct higher levels of output voltage. We applied four SIIs from 9-level to 49-level of output voltage to a solar panel energy system and obtained the satisfactory simulation and experimental results, which strongly supported our

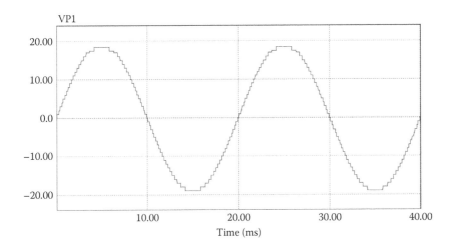

FIGURE 13.12
A 39-level simulation waveform.

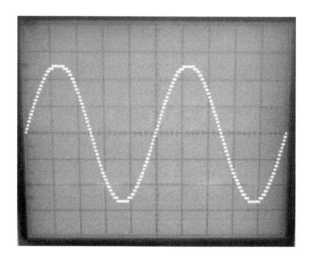

FIGURE 13.13
A 39-level experimental waveform.

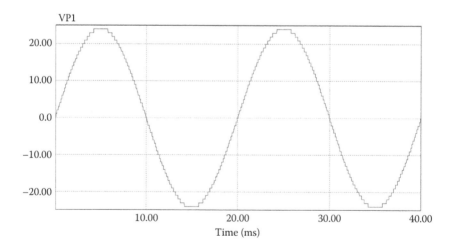

FIGURE 13.14
A 49-level simulation waveform.

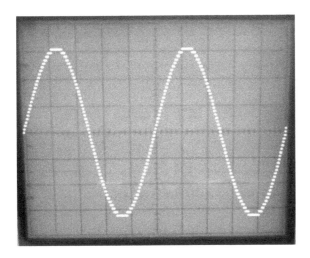

FIGURE 13.15
A 49-level experimental waveform.

circuit design. These SI multilevel inverters can be used in other renewable energy systems and industrial applications.

References

1. Luo F. L. and Ye H. 2010. *Power Electronics: Advanced Conversion Technologies.* Boca Raton, FL: Taylor & Francis.
2. Luo F. L. and Ye H. 2004. *Advanced DC/DC Converters.* Boca Raton, FL: CRC Press.
3. Luo F. L. and Ye H. 2000. Multi-quadrant switched inductor luo-converter. *Int. J. Power Supply Technol. Applicat.* Vol. 3, No. 6, pp. 258–263.

14

Best Switching Angles to Obtain Lowest THD for Multilevel DC/AC Inverters

The lowest THD of multilevel DC/AC inverters from 3-level to 81-level is derived in this chapter. This kind of multilevel DC/AC inverter can be applied in renewable energy systems, electrical vehicles, and other industrial applications [1,2].

14.1 Introduction

Multilevel DC/AC inverters have various structures and many advantages. Unfortunately, most existing inverters are unable to produce good output AC waveforms because of their poor total harmonic distortion (THD) because each level switching angle is not carefully arranged. In order to gain good power quality (PQ), we have to carefully investigate the switching angle arrangement to obtain the lowest THD.

14.2 Methods for Determination of Switching Angle

Switching angle is the moment of the level change. Referring to Figure 14.1, for an m-level (m is an odd number) waveform in the period $0°$–$90°$, there are $2(m - 1)$ switching angles to be determined. We define them as α_1, α_2, ... α_{m-2}, α_{m-1} by the time sequence. Since the sine wave is a symmetrical wave, the negative half-cycle is centrally symmetrical to its positive half cycle; the wave of the second quarter period is mirror-symmetrical to its positive half-cycle; and the wave of the second quarter period is mirror-symmetrical to the wave of its first quadrant period. We define the switching angles in the first quadrant period (i.e., $0°$–$90°$) as main switching angles.

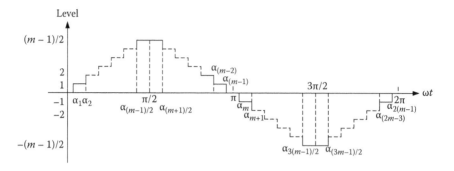

FIGURE 14.1
Output voltage waveform for multilevel inverter.

14.2.1 Main Switching Angles

For an m-level (m is an odd number) waveform, there are $(m - 1)/2$ main switching angles. Referring to Figure 14.1, we have the following relations:

1. Main switching angles in the first quadrant (i.e., 0°–90°):

   $\alpha_1, \alpha_2, \ldots, \alpha_{(m-1)/2}.$

2. The switching angles in the second quadrant (i.e., 90°–180°):

   $\alpha_{(m+1)/2} = \pi - \alpha_{(m-1)/2}, \ldots, \alpha_{(m-1)} = \pi - \alpha_1.$

3. The switching angles in the third quadrant (i.e., 180°–270°): $= \pi - \alpha$

   $\alpha_m = \pi + \alpha_1, \ldots, \alpha_{3(m-1)/2} = \pi + \alpha_{(m-1)/2}.$

4. The switching angles in the fourth quadrant (i.e., 270°–360°):

   $\alpha_{(3m-1)/2} = 2\pi - \alpha_{(m-1)/2}, \ldots, \alpha_{2(m-1)} = 2\pi - \alpha_1.$

From the analysis, we need only determine the main switching angles. The other switching angles can be derived from the main switching angles in the first quadrant (i.e., 0°–90°):

$$\alpha_1, \alpha_2, \ldots, \alpha_{(m-1)/2}.$$

14.2.2 Equal-Phase (EP) Method

The equal-phase (EP) method is derived from the simplest idea, to averagely distribute the switching angles in the range 0–π. The main switching angles are determined by the formula:

$$\alpha_i = i\frac{180°}{m} \quad \text{where } i = 1, 2, \ldots \frac{m-1}{2} \tag{14.1}$$

14.2.3 Half-Equal-Phase (HEP) Method

Since the multilevel waveform determined by the EP method looks very narrow, like a triangle waveform, another approach called the half-equal-phase (HEP) method to arrange the main switching angles can obtain a wider and better output waveform. The main switching angles are in the range $0–\pi/2$, which are determined by the formula:

$$\alpha_i = i \cdot \frac{90°}{\frac{m+1}{2}} = i \cdot \frac{180°}{m+1} \quad where \quad i = 1, 2, \ldots \frac{m-1}{2} \tag{14.2}$$

14.2.4 Half-Height (HH) Method

The above two methods are able to arrange the main switching angles in a simple manner, but the output waveform is not a sine wave. According to the sine function, we established a new half-height (HH) method to determine the main switching angles. The idea is that when the function value increases to the half-height of the level, the switch angle is set and thus a better output waveform obtained. The main switching angles are determined by the formula:

$$\alpha_i = \sin^{-1}\left[\left(i-\frac{1}{2}\right)\frac{2}{m-1}\right] = \sin^{-1}\left(\frac{2i-1}{m-1}\right) where \ i = 1, 2, \ldots \frac{m-1}{2} \tag{14.3}$$

14.2.5 Feed-Forward (FF) Method

Using the above three methods, we can see that there are wider gaps between the positive half-cycle and the negative half-cycle. In order to reduce the gaps, we established another new method, the feed-forward (FF) method, to determine the main switching angles by the formula:

$$\alpha_i = \frac{1}{2}\sin^{-1}\left[\left(i-\frac{1}{2}\right)\frac{2}{m-1}\right] = \frac{1}{2}\sin^{-1}\left(\frac{2i-1}{m-1}\right) where \ i = 1, 2, \ldots \frac{m-1}{2} \tag{14.4}$$

14.2.6 Comparison of Methods in Each Level

For $m = 3$, we have only one main switching angle α_1. We compare them in Table 14.1.

For $m = 5$, we have two main switching angles α_1 and α_2 in Table 14.2.

For $m = 7$, we have three main switching angles α_1, α_2, and α_3 in Table 14.3.

For $m = 9$, we have four main switching angles α_1, α_2, α_3, and α_4 in Table 14.4.

For $m = 11$, we have five main switching angles α_1, α_2, α_3, α_4 and α_5 in Table 14.5.

For $m = 13$, we have six main switching angles α_1, α_2, α_3, α_4, α_5 and α_6 in Table 14.6.

TABLE 14.1

Comparison of Switching Angle α_1 of the Methods (m = 3)

Methods	Switching Angle α_1(°)	THD
EP	60°	80.17%
HEP	45°	48.19%
HH	30°	30.9%
FF	15°	31.76%

TABLE 14.2

Comparison of Switching Angles of the Methods (m = 5)

Methods	Switching Angle α_1(°)	Switching Angle α_2(°)	THD
EP	36°	72°	42.77%
HEP	30°	60°	31.78%
HH	14.48°	49°	21.14%
FF	7.24°	24.5°	24.86%

TABLE 14.3

Comparison of Switching Angles of the Methods (m = 7)

Methods	α_1(°)	α_2(°)	α_3(°)	THD
EP	25.71°	51.43°	77.14°	30.98%
HEP	22.50	45.00	67.50	31.29%
HH	9.60	30.00	56.44	11.70%
FF	4.80	15.00	28.22	22.17%

TABLE 14.4

Comparison of Switching Angles of the Methods (m = 9)

Methods	α_1(°)	α_2(°)	α_3(°)	α_4(°)	THD
EP	20.00	40.00	60.00	80.00	25.37%
HEP	18.00	36.00	54.00	72.00	22.06%
HH	7.20	22.00	38.70	61.10	8.37%
FF	3.60	11.00	19.30	30.50	21.30%

TABLE 14.5

Comparison of Switching Angles of the Methods (m = 11)

Methods	$\alpha_1(°)$	$\alpha_2(°)$	$\alpha_3(°)$	$\alpha_4(°)$	$\alpha_5(°)$	THD
EP	16.36	32.72	49.09	65.45	81.81	22.62%
HEP	15.00	30.00	45.00	60.00	75.00	20.16%
HH	5.74	17.46	30.00	44.43	64.16	7.72%
FF	2.87	8.73	15.00	22.21	32.08	21.24%

TABLE 14.6

Comparison of Switching Angles of the Methods (m = 13)

Methods	$\alpha_1(°)$	$\alpha_2(°)$	$\alpha_3(°)$	$\alpha_4(°)$	$\alpha_5(°)$	$\alpha_6(°)$	THD
EP	13.85	27.69	41.54	55.38	69.23	83.08	20.25%
HEP	12.86	25.71	38.57	51.43	64.29	77.14	18.74%
HH	4.78	14.48	24.62	35.69	48.59	66.44	7.25%
FF	2.39	7.24	12.31	17.85	24.29	33.22	21.00%

For *m* = 15, we have seven main switching angles $\alpha_1, \alpha_2, \alpha_3, \alpha_4, \alpha_5, \alpha_6$ and α_7 in Table 14.7.

High number of m can be listed accordingly. The different levels for each method are given in the next section.

14.2.7 Comparison of Levels for Each Method

We compare the various levels for each method to find out which method is better to obtain lower THD (Tables 14.8 to 14.11).

14.2.8 THDs of Different Methods

Comparisons of THDs for the different methods are listed in Table 14.12.

TABLE 14.7

Comparison of Switching Angles of the Methods (m = 15)

Methods	$\alpha_1(°)$	$\alpha_2(°)$	$\alpha_3(°)$	$\alpha_4(°)$	$\alpha_5(°)$	$\alpha_6(°)$	$\alpha_7(°)$	THD
EP	12.0	24.00	36.00	48.00	60.00	72.00	84.00	18.56%
HEP	11.25	22.50	33.75	45.00	56.25	67.50	78.75	17.85%
HH	4.10	12.37	20.92	30.00	40.01	51.79	68.21	5.67%
FF	2.05	6.19	10.46	15.00	20.00	25.90	34.11	20.92%

TABLE 14.8

THD of Different Levels Using EP Method with $m = 35$

THD	80.17%	42.77%	30.98%	25.37%	22.62%	20.25%	18.56%	17.55 %%%%%%	17.20%	16.48%	16.15%	15.78%	15.33%	15.02%	14.60%	14.26%	13.93%
α_{17} (°)																	87.38
α_{16} (°)																87.20	82.24
α_{15} (°)															87.10	81.75	77.10
α_{14} (°)														86.90	81.29	76.30	71.96
α_{13} (°)													86.67	80.69	75.48	70.85	66.82
α_{12} (°)												86.40	80.00	74.48	69.68	65.40	61.68
α_{11} (°)											86.09	79.20	73.33	68.28	63.87	59.95	56.54
α_{10} (°)										85.71	78.26	72.00	66.67	62.07	58.06	54.50	51.40
α_{9} (°)									85.26	77.14	70.43	64.80	60.00	55.86	52.26	49.05	46.26
α_{8} (°)								84.71	75.79	68.57	62.61	57.60	53.33	49.66	46.45	43.60	41.12
α_{7} (°)							84.00	74.12	66.32	60.00	54.78	50.40	46.67	43.45	40.65	38.15	35.98
α_{6} (°)						83.08	72.00	63.53	56.84	51.43	46.96	43.20	40.00	37.243 / 43.45	34.84	32.70	30.84
α_{5} (°)					81.81	69.23	60.00	52.94	47.37	42.86	39.13	36.00	33.33	31.03	29.03	27.25	25.70
α_{4} (°)				80.00	65.45	55.38	48.00	42.35	37.89	34.29	31.30	28.8	26.67	24.83	23.23	21.8	20.56
α_{3} (°)			77.14	60.00	49.09	41.54	36.00	31.76	28.42	25.71	23.49	21.6	20.00	18.62	17.42	16.35	15.42
α_{2} (°)		72.00	51.43	40.00	32.722	27.69	24.00	21.20	18.90	17.14	15.70	14.40	13.33	12.41	11.61	10.90	10.38
α_{1} (°)	60.00	36.00	25.71	20.00	16.36	13.85	12.00	10.59	9.47	8.57	7.83	7.20	6.67	6.21	5.81	5.45	5.14
level	3	5	7	9	11	13	15	17	19	21	23	25	27	29	31	33	35

TABLE 14.9

THD of Different Levels Using HEP Method with m = 35

THD	48.19%	31.78%	31.29%	22.06%	20.16%	18.74%	17.85%	16.35%	16.30%	16.12%	15.44%	14.90%	14.59%	14.12%	13.79%	13.35%	12.99%
$\alpha_{17}(°)$																	85.00
$\alpha_{16}(°)$																84.71	80.00
$\alpha_{15}(°)$															84.38	79.41	75.00
$\alpha_{14}(°)$														84.00	78.75	74.12	70.00
$\alpha_{13}(°)$													83.59	78.00	73.13	68.82	65.00
$\alpha_{12}(°)$												83.08	77.16	72.00	67.5	63.53	60.00
$\alpha_{11}(°)$											82.50	76.15	70.73	66.00	61.808	58.24	55.00
$\alpha_{10}(°)$										81.80	75.00	69.23	64.30	60.00	56.25	52.904	50.00
$\alpha_{9}(°)$									81.00	73.60	67.50	62.31	57.87	54.00	50.63	47.65	45.00
$\alpha_{8}(°)$								80.00	72.00	65.24	60.00	55.38	51.44	48.00	45.00	42.35	40.00
$\alpha_{7}(°)$							78.75	70.00	63.00	57.27	52.50	48.16	45.01	42.00	39.38	37.06	35.00
$\alpha_{6}(°)$						77.14	67.50	60.00	54.00	49.09	45.00	41.54	38.58	36.00	33.75	31.76	30.00
$\alpha_{5}(°)$					75.00	64.29	56.25	50.00	45.00	40.91	37.50	34.62	32.15	30.00	28.13	26.47	25.00
$\alpha_{4}(°)$				72.00	60.00	51.43	45.00	40.00	36.00	32.73	30.00	27.69	25.72	24.00	22.50	21.18	20.00
$\alpha_{3}(°)$			67.50	54.00	45.00	38.57	33.75	30.00	27.00	24.55	22.50	20.77	19.29	18.00	16.88	15.88	15.00
$\alpha_{2}(°)$		60.00	45.00	36.00	30.00	25.71	22.50	20.00	18.00	16.36	15.00	13.85	12.86	12.00	11.25	10.59	10.00
$\alpha_{1}(°)$	45.00	30.00	22.50	18.00	15.00	12.86	11.25	10.00	9.00	8.18	7.50	6.92	6.43	6.00	5.63	5.29	5.00
level	3	5	7	9	11	13	15	17	19	21	23	25	27	29	31	33	35

TABLE 14.10

THD of Different Levels Using HH Method with m = 35

THD	30.90%	21.14 %%%%	11.70%	8.37%	7.72%	7.25%	5.67%	5.02%	4.75%	4.70%	4.68%	4.63%	4.58%	4.51%	4.46%	4.39%	4.36%
$\alpha_{17}(°)$																	76.00
$\alpha_{16}(°)$																75.64	65.75
$\alpha_{15}(°)$															75.16	64.99	58.53
$\alpha_{14}(°)$														74.64	64.16	57.54	52.57
$\alpha_{13}(°)$													74.06	63.23	56.44	51.38	47.33
$\alpha_{12}(°)$												73.40	62.20	55.23	50.06	45.95	42.57
$\alpha_{11}(°)$											72.66	61.04	53.87	48.59	44.43	41.01	38.14
$\alpha_{10}(°)$										71.81	59.73	52.34	46.95	42.73	39.30	36.42	33.97
$\alpha_{9}(°)$									70.81	58.21	50.6	45.10	40.85	37.38	34.52	32.09	30.00
$\alpha_{8}(°)$								69.64	56.44	48.59	42.99	38.68	35.23	32.39	30.00	27.95	26.18
$\alpha_{7}(°)$							68.21	54.34	46.24	40.54	36.22	32.80	30.00	27.66	25.68	23.97	22.48
$\alpha_{6}(°)$						66.44	51.79	43.43	37.67	33.37	30.00	27.28	25.03	23.13	21.51	20.11	18.88
$\alpha_{5}(°)$					64.16	48.59	40.01	34.23	30.00	26.74	24.15	22.02	20.25	18.75	17.46	16.33	15.35
$\alpha_{4}(°)$				61.10	44.43	35.69	30.00	25.94	22.89	20.49	18.55	16.96	15.62	14.48	13.49	12.64	11.88
$\alpha_{3}(°)$			56.44	38.70	30.00	24.62	20.92	18.21	16.13	14.48	13.14	12.02	11.09	10.29	9.59	8.99	8.46
$\alpha_{2}(°)$		49.00	30.00	22.00	17.46	14.48	12.37	10.81	9.59	8.63	7.84	7.18	6.63	6.15	5.74	5.38	5.06
$\alpha_{1}(°)$	30.00	14.48	9.60	7.20	5.74	4.78	4.10	3.58	3.18	2.87	2.61	2.39	2.20	2.05	1.91	1.79	1.69
level	3	5	7	9	11	13	15	17	19	21	23	25	27	29	31	33	35

TABLE 14.11

THD of Different Levels Using FFM Method

THD	31.76%	24.86%	22.17%	21.10%	21.24%	21.00%	20.92%	20.90%	20.80%	20.76%	20.74%	20.67%	20.64%	20.59%	20.53%	20.46%	20.44%
$\alpha_{17}(°)$																	38.00
$\alpha_{16}(°)$																37.82	32.88
$\alpha_{15}(°)$															37.58	32.50	29.27
$\alpha_{14}(°)$														37.32	32.08	28.77	26.29
$\alpha_{13}(°)$													37.03	31.62	28.22	25.69	23.67
$\alpha_{12}(°)$												36.70	31.10	27.62	25.03	22.98	21.28
$\alpha_{11}(°)$											36.33	30.52	26.94	24.30	22.22	20.51	19.07
$\alpha_{10}(°)$										35.91	29.87	26.17	23.48	21.37	19.65	18.21	16.99
$\alpha_{9}(°)$									35.40	2911	25.3	22.55	20.43	18.69	17.26	16.05	15.00
$\alpha_{8}(°)$								34.82	28.22	24.30	21.50	19.34	17.62	16.20	15.00	13.98	13.09
$\alpha_{7}(°)$							34.11	27.17	23.12	20.27	18.11	16.40	15.00	13.83	12.84	11.99	11.24
$\alpha_{6}(°)$						33.20	25.90	21.72	18.84	16.69	15.00	13.64	12.01	11.57	10.76	10.06	9.44
$\alpha_{5}(°)$					32.008	24.29	20.00	17.12	15.00	13.37	12.08	11.01	10.13	9.38	8.73	8.17	7.68
$\alpha_{4}(°)$				30.50	22.21	17.85	15.00	12.97	11.45	10.25	9.28	8.48	7.81	7.24	6.75	6.32	5.94
$\alpha_{3}(°)$			28.22	19.30	15.00	12.31	10.46	9.11	8.07	7.24	6.57	6.01	5.55	5.15	4.80	4.50	4.23
$\alpha_{2}(°)$		4.24	15.00	11.00	8.73	7.24	6.19	5.41	4.80	4.32	3.92	3.59	3.32	3.08	2.87	2.69	2.53
$\alpha_{1}(°)$	15.00	24.50	4.80	3.60	2.87	2.39	2.05	1.79	1.59	1.44	1.31	1.20	1.10	1.03	0.96	0.90	0.85
level	3	5	7	9	11	13	15	17	19	21	23	25	27	29	31	33	35

TABLE 14.12

THD of Different Methods

Level	EPM	HEPM	HHM	FFM	Best Firing Angle
3	80.17%	48.19%	30.90%	31.76%	28.87%
5	42.77%	31.78%	21.14%	24.86%	16.42%
7	30.98%	31.29%	11.70%	22.17%	11.53%
9	25.37%	22.06%	8.37%	21.30%	8.90%
11	22.62%	20.16%	7.72%	21.24%	7.26%
13	20.25%	18.74%	7.25%	21.00%	6.13%
15	18.56%	17.85%	5.67%	20.92%	5.31%
17	17.55%	16.35%	5.02%	20.90%	4.68%
19	17.20%	16.44%	4.75%	20.80%	4.19%
21	16.48%	16.12%	4.70%	20.76%	3.79%
23	16.15%	15.44%	4.68%	20.74%	3.46%
25	15.78%	14.90%	4.63%	20.67%	3.18%
27	15.33%	14.59%	4.58%	20.64%	2.95%
29	15.02%	14.12%	4.51%	20.59%	2.74%
31	14.60%	13.79%	4.46%	20.53%	2.57%
33	14.26%	13.35%	4.39%	20.46%	2.41%
35	13.93%	12.99%	4.36%	20.44%	2.28%

14.3 Best Switching Angles

From Tables 14.1 to 14.12, we can see that THD is reduced when the number of levels (m) of the inverter increases, and the HH method is better than the other three methods. Hence, a higher level of inverter will be considered to produce output with less harmonic content.

14.3.1 Using MATLAB to Obtain Best Switching Angles

We use MATLAB software to search for the best switching angles in this section, and results (for $m = 81$) are shown in Table 14.13.

14.3.2 Analysis of Results of Best Switching Angles Calculation

THD values obtained using best switching angles from Table 14.13 are listed below.

From Table 14.14, the lowest THD value of a multilevel inverter with level equal to or below 81 is 0.99%. It can be easily observed that the differences between each adjacent level decrease gradually as the number of levels increase. For example, the THD value drops by 12.54% when the number of level increases from 3 to 5. However, the THD value drops by only 0.02% when the number of level increases from 79 to 81. By applying the MATLAB

TABLE 14.13

Best Switching Angles

Level	α1(rad)	α2(rad)	α3(rad)	α4(rad)	α5(rad)	α6(rad)	α7(rad)	α8(rad)	α9(rad)	α10(rad)	α11(rad)	α12(rad)	α13(rad)	α14(rad)	α15(rad)	α16(rad)	α17(rad)	α18(rad)	α19(rad)	α20(rad)
3	0.4053																			
5	0.2242	0.7301																		
7	0.1550	0.4817	0.8821																	
9	0.1185	0.3625	0.6323	0.9744																
11	0.0958	0.2912	0.4989	0.7341	1.3078															
13	0.0804	0.2436	0.4136	0.5976	0.8088	1.0848														
15	0.0693	0.2094	0.3538	0.5064	0.6733	0.8666	1.1214													
17	0.0609	0.1836	0.3093	0.4402	0.5798	0.7337	0.9130	1.1509												
19	0.0544	0.1635	0.2750	0.3897	0.5105	0.6400	0.7834	0.9513	1.1754											
21	0.0490	0.1475	0.2474	0.3500	0.4565	0.5690	0.6902	0.8252	0.9839	1.1961										
23	0.0446	0.1342	0.2250	0.3176	0.4132	0.5129	0.6187	0.7332	0.8610	1.0117	1.2138									
25	0.4090	0.1232	0.2063	0.2909	0.3776	0.4674	0.5616	0.6619	0.7704	0.8921	1.0358	1.2292								
27	0.0379	0.1139	0.1905	0.2682	0.3477	0.4296	0.5147	0.6041	0.6994	0.8031	0.9194	1.0572	1.2429							
29	0.0352	0.1058	0.1769	0.2490	0.3224	0.3976	0.4754	0.5564	0.6416	0.7328	0.8321	0.9438	1.0762	1.2550						
31	0.0325	0.0989	0.1652	0.2324	0.3006	0.3703	0.4419	0.5160	0.5934	0.6751	0.7626	0.8580	0.9656	1.0932	1.2661					
33	0.0309	0.0927	0.1550	0.2178	0.2815	0.3464	0.4128	0.4813	0.5525	0.6267	0.7052	0.7894	0.8815	0.9852	1.1088	1.2760				
35	0.0291	0.0874	0.1458	0.2050	0.2648	0.3256	0.3873	0.4511	0.5170	0.5850	0.6566	0.7323	0.8136	0.9024	1.0032	1.1226	1.2846			
37	0.0276	0.0824	0.1377	0.1936	0.2501	0.3070	0.3652	0.4249	0.4860	0.5493	0.6149	0.6842	0.7573	0.8359	0.9220	1.0194	1.1353	1.2932		
39	0.0261	0.0781	0.1307	0.1834	0.2366	0.2907	0.3455	0.4014	0.4587	0.5177	0.5787	0.6424	0.7092	0.7800	0.8562	0.9399	1.0344	1.1473	1.3006	
41	0.0246	0.0745	0.1234	0.1740	0.2247	0.2759	0.3276	0.3802	0.4343	0.4897	0.5468	0.6056	0.6672	0.7318	0.8009	0.8747	0.9560	1.0484	1.1579	1.3071
43	0.0236	0.0709	0.1183	0.1658	0.2139	0.2625	0.3117	0.3616	0.4127	0.4647	0.5182	0.5735	0.6307	0.6905	0.7534	0.8201	0.8921	0.9713	1.0609	1.1682
45	0.0225	0.0676	0.1128	0.1583	0.2041	0.2504	0.2972	0.3447	0.3931	0.4423	0.4929	0.5446	0.5981	0.6539	0.7119	0.7730	0.8381	0.9082	0.9854	1.0729
47	0.0221	0.0652	0.1079	0.1515	0.1952	0.2395	0.2846	0.3289	0.3751	0.4216	0.4700	0.5187	0.5692	0.6213	0.6752	0.7319	0.7916	0.8548	0.9236	0.9983
49	0.0206	0.0621	0.1034	0.1451	0.1870	0.2294	0.2721	0.3152	0.3589	0.4035	0.4490	0.4954	0.5431	0.5921	0.6427	0.6955	0.7507	0.8086	0.8707	0.9376
51	0.0197	0.0595	0.0995	0.1394	0.1795	0.2201	0.2611	0.3023	0.3442	0.3867	0.4298	0.4740	0.5192	0.5656	0.6133	0.6627	0.7142	0.7679	0.8245	0.8852
53	0.0186	0.0550	0.0921	0.1292	0.1660	0.2035	0.2413	0.2795	0.3178	0.3568	0.3962	0.4366	0.4775	0.5194	0.5625	0.6065	0.6518	0.6991	0.7481	0.7996
55	0.0186	0.0550	0.0921	0.1292	0.1660	0.2035	0.2413	0.2795	0.3178	0.3568	0.3962	0.4366	0.4775	0.5194	0.5625	0.6065	0.6518	0.6991	0.7481	0.7996

(continued)

TABLE 14.13 (CONTINUED)

Best Switching Angles

Level	α1(rad)	α2(rad)	α3(rad)	α4(rad)	α5(rad)	α6(rad)	α7(rad)	α8(rad)	α9(rad)	α10(rad)	α11(rad)	α12(rad)	α13(rad)	α14(rad)	α15(rad)	α16(rad)	α17(rad)	α18(rad)	α19(rad)	α20(rad)
57	0.0177	0.0533	0.0888	0.1244	0.1603	0.1963	0.2327	0.2693	0.3063	0.3436	0.3817	0.4202	0.4595	0.4993	0.5402	0.5822	0.6253	0.6698	0.7158	0.7639
59	0.0171	0.0514	0.0859	0.1202	0.1548	0.1896	0.2246	0.2598	0.2953	0.3318	0.3681	0.4051	0.4426	0.4807	0.5198	0.5597	0.6008	0.6431	0.6864	0.7314
61	0.0166	0.0498	0.0830	0.1162	0.1494	0.1832	0.2169	0.2509	0.2856	0.3202	0.3551	0.3909	0.4270	0.4635	0.5009	0.5392	0.5782	0.6183	0.6595	0.7022
63	0.0160	0.0481	0.0802	0.1125	0.1448	0.1773	0.2100	0.2429	0.2761	0.3095	0.3434	0.3776	0.4124	0.4477	0.4835	0.5201	0.5575	0.5957	0.6350	0.6754
65	0.0155	0.0466	0.0778	0.1090	0.1402	0.1717	0.2033	0.2352	0.2673	0.2996	0.3323	0.3654	0.3988	0.4328	0.4673	0.5024	0.5382	0.5748	0.6123	0.6508
67	0.0150	0.0452	0.0754	0.1056	0.1360	0.1665	0.1972	0.2280	0.2590	0.2903	0.3219	0.3538	0.3862	0.4189	0.4521	0.4859	0.5203	0.5554	0.5913	0.6280
69	0.0146	0.0439	0.0732	0.1025	0.1320	0.1616	0.1913	0.2212	0.2513	0.2816	0.3122	0.3431	0.3743	0.4059	0.4380	0.4705	0.5037	0.5374	0.5718	0.6070
71	0.0142	0.0426	0.0711	0.0996	0.1282	0.1570	0.1857	0.2148	0.2440	0.2734	0.3030	0.3330	0.3632	0.3937	0.4247	0.4562	0.4881	0.5205	0.5536	0.5874
73	0.0139	0.0415	0.0691	0.0969	0.1247	0.1526	0.1806	0.2088	0.2371	0.2656	0.2944	0.3234	0.3527	0.3823	0.4123	0.4427	0.4735	0.5048	0.5366	0.5691
75	0.0136	0.0403	0.0673	0.0943	0.1213	0.1485	0.1757	0.2031	0.2306	0.2583	0.2863	0.3144	0.3428	0.3715	0.4005	0.4299	0.4597	0.4900	0.5207	0.5519
77	0.0133	0.0392	0.0655	0.0918	0.1180	0.1446	0.1711	0.1978	0.2245	0.2515	0.2785	0.3060	0.3335	0.3612	0.3896	0.4182	0.4468	0.4763	0.5056	0.5362
79	0.0129	0.0380	0.0638	0.0894	0.1153	0.1408	0.1666	0.1926	0.2187	0.2448	0.2712	0.2979	0.3247	0.3518	0.3792	0.4065	0.4345	0.4629	0.4917	0.5209
81	0.0125	0.0373	0.0622	0.0872	0.1123	0.1373	0.1625	0.1877	0.2132	0.2386	0.2643	0.2903	0.3163	0.3426	0.3691	0.3960	0.4231	0.4505	0.4784	0.5066

Level	α21(rad)	α22(rad)	α23(rad)	α24(rad)	α25(rad)	α26(rad)	α27(rad)	α28(rad)	α29(rad)	α30(rad)	α31(rad)	α32(rad)	α33(rad)	α34(rad)	α35(rad)	α36(rad)	α37(rad)	α38(rad)	α39(rad)	α40(rad)
43	1.3138																			
45	1.1775	1.3199																		
47	1.0841	1.1867	1.3253																	
49	1.0112	1.0948	1.1949	1.3313																
51	0.9506	1.0224	1.1043	1.2022	1.3357															
53	0.8990	0.9630	1.0333	1.1135	1.2092	1.3405														
55	0.8539	0.9118	0.9743	1.0433	1.1220	1.2158	1.3443													
57	0.8143	0.8674	0.9242	0.9855	1.0532	1.1303	1.2227	1.3490												
59	0.7786	0.8280	0.8801	0.9360	0.9959	1.0627	1.1381	1.2289	1.3530											
61	0.7466	0.7925	0.8412	0.8923	0.9466	1.0058	1.0711	1.1455	1.2345	1.3567										
63	0.7172	0.7606	0.8059	0.8533	0.9035	0.9572	1.0153	1.0795	1.1525	1.2402	1.3602									
65	0.6904	0.7314	0.7740	0.8185	0.8651	0.9144	0.9671	1.0243	1.0873	1.1592	1.2454	1.3635								
67	0.6658	0.7047	0.7450	0.7868	0.8304	0.8763	0.9248	0.9766	1.0328	1.0949	1.1656	1.2505	1.3668							
69	0.6430	0.6801	0.7183	0.7579	0.7989	0.8418	0.8869	0.9346	0.9856	1.0409	1.1020	1.1716	1.2553	1.3699						
71	0.6219	0.6573	0.6938	0.7313	0.7702	0.8106	0.8528	0.8971	0.9441	0.9943	1.0487	1.1089	1.1774	1.2598	1.3728					
73	0.6022	0.6362	0.6710	0.7068	0.7437	0.7819	0.8217	0.8632	0.9069	0.9531	1.0025	1.0562	1.1155	1.1830	1.2642	1.3756				
75	0.5839	0.6165	0.6498	0.6840	0.7192	0.7555	0.7932	0.8323	0.8732	0.9161	0.9617	1.0104	1.0632	1.1216	1.1882	1.2683	1.3781			
77	0.5666	0.5980	0.6302	0.6631	0.6965	0.7316	0.7671	0.8041	0.8423	0.8827	0.9253	0.9700	1.0179	1.0703	1.1278	1.1937	1.2728	1.3813		
79	0.5505	0.5809	0.6115	0.6433	0.6755	0.7088	0.7428	0.7778	0.8143	0.8524	0.8921	0.9338	0.9783	1.0252	1.0768	1.1337	1.1986	1.2766	1.3836	
81	0.5354	0.5646	0.5944	0.6246	0.6556	0.6873	0.7201	0.7537	0.7883	0.8244	0.8618	0.9009	0.9421	0.9856	1.0324	1.0830	1.1392	1.2031	1.2802	1.3858

TABLE 14.14

THD Value Obtained Using Best Switching Angles

Number of Levels	THD	Difference from Lower Levels
3	28.96%	
5	16.42%	12.54%
7	11.53%	4.89%
9	8.90%	2.63%
11	7.26%	1.64%
13	6.13%	1.13%
15	5.31%	0.82%
17	4.68%	0.63%
19	4.19%	0.49%
21	3.79%	0.40%
23	3.46%	0.33%
25	3.18%	0.28%
27	2.95%	0.23%
29	2.74%	0.21%
31	2.57%	0.17%
33	2.41%	0.16%
35	2.28%	0.13%
37	2.15%	0.13%
39	2.04%	0.11%
41	1.94%	0.10%
43	1.85%	0.09%
45	1.77%	0.08%
47	1.70%	0.07%
49	1.63%	0.07%
51	1.56%	0.07%
53	1.51%	0.05%
55	1.45%	0.06%
57	1.40%	0.05%
59	1.35%	0.05%
61	1.31%	0.04%
63	1.27%	0.04%
65	1.23%	0.04%
67	1.19%	0.04%
69	1.16%	0.03%
71	1.13%	0.03%
73	1.10%	0.03%
75	1.07%	0.03%
77	1.04%	0.03%
79	1.01%	0.03%
81	0.99%	0.02%

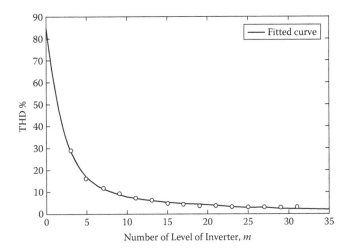

FIGURE 14.2
THD versus *m*.

graph fitting tool, the relationship between lowest THD and number of levels of an inverter can be shown as the following equation:

$$THD_{Lowest} = 72.42e^{-0.4503m} + 11.86e^{-0.05273m} \tag{14.5}$$

where *m* is the level number of the inverter. The corresponding figure for THD versus *m* is shown in Figure 14.2.

14.3.3 Output Voltage Waveform for Multilevel Inverters

To verify our design, simulation results of the output voltage waveform for multilevel inverters with levels from 7 to 35 are shown in this section. See Figures 14.3 to 14.17.

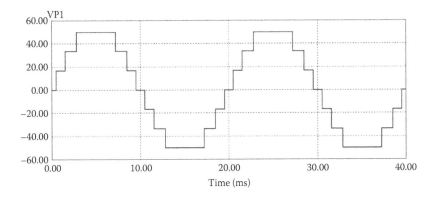

FIGURE 14.3
Output voltage waveform of seven-level inverter.

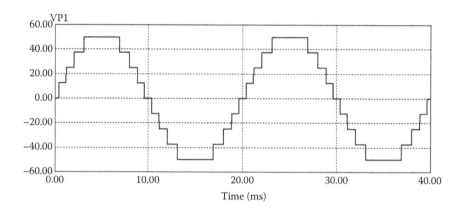

FIGURE 14.4
Output voltage waveform of nine-level inverter.

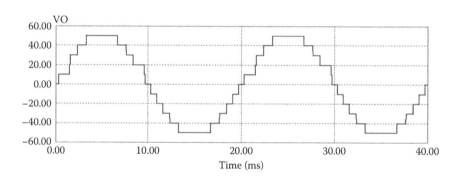

FIGURE 14.5
Output voltage waveform of 11-level inverter.

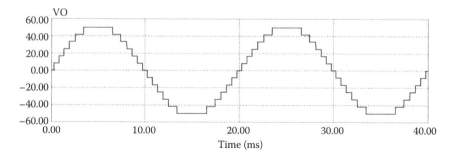

FIGURE 14.6
Output voltage waveform of 13-level inverter.

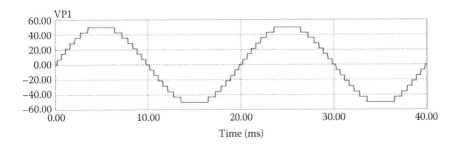

FIGURE 14.7
Output voltage waveform of 15-level inverter.

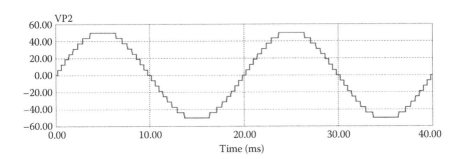

FIGURE 14.8
Output voltage waveform of 17-level inverter.

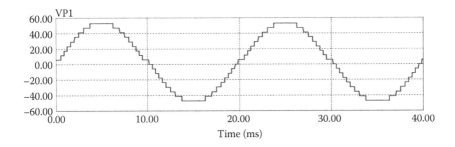

FIGURE 14.9
Output voltage waveform of 19-level inverter.

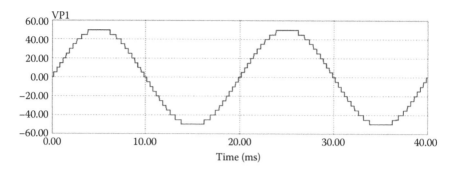

FIGURE 14.10
Output voltage waveform of 21-level inverter.

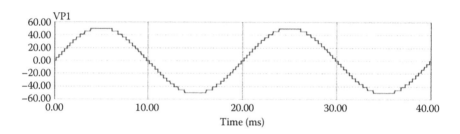

FIGURE 14.11
Output voltage waveform of 23-level inverter.

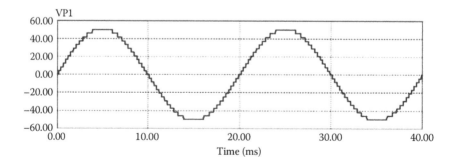

FIGURE 14.12
Output voltage waveform of 25-level inverter.

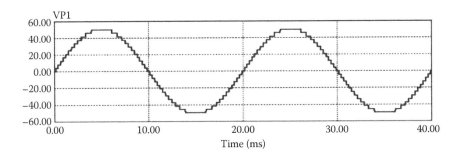

FIGURE 14.13
Output voltage waveform of 27-level inverter.

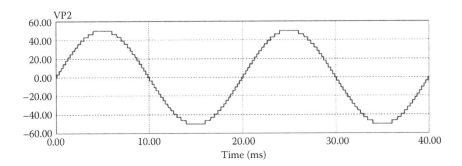

FIGURE 14.14
Output Voltage Waveform of 29-level Inverter

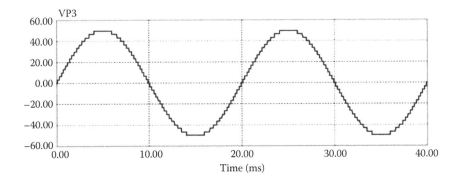

FIGURE 14.15
Output voltage waveform of 31-level inverter.

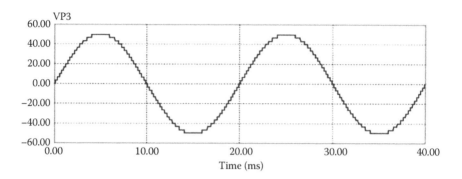

FIGURE 14.16
Output voltage waveform of 33-level inverter.

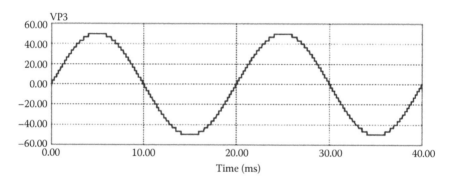

FIGURE 14.17
Output voltage waveform of 35-level inverter.

References

1. Luo F. L. and Ye H. 2010. *Power Electronics: Advanced Conversion Technologies.* Boca Raton, FL: Taylor & Francis.
2. Fang Lin Luo. 2012. Best Switching Angles to Obtain Lowest THD for Multilevel DC/AC Inverters. NTU Technical Report.

15

Design Examples for Wind Turbine and Solar Panel Energy Systems

Wind turbine and solar panel energy are clean and renewable. In recent years, their applications have attracted much worldwide attention. Therefore, the design of wind turbine and solar panel energy systems is a very popular research area.

15.1 Introduction

We first introduce some units used to measure large values of power. They are grouped by order of magnitude as follows:

- kW—kilowatt (10^3 W)
- MW—megawatt (10^6 W)
- GW—gigawatt (10^9 W)
- TW—terawatt (10^{12} W)
- PW—petawatt (10^{15} W)
- EW—exawatt (10^{18} W)
- ZW—zettawatt (10^{21} W)
- YW—yottawatt (10^{24} W)

The relationship between the watt and joule is: 1 joule = 1 watt × 1 second.

The sun radiates 3.8×10^{20} MW into space. Earth receives 174 petawatts (PW) of incoming solar radiation (insolation) at the upper atmosphere, as shown in Figure 15.1 [1,2]. Approximately 30% of the power from the sun is reflected back to outer space, while the rest is absorbed by air, clouds, oceans, and land masses. The spectrum of solar light at the surface of Earth is mostly spread across the visible and near-infrared ranges with a small part in the near-ultraviolet range.

Earth's land surface, oceans, and atmosphere absorb solar radiation, and their temperature rises. Warm air contains evaporated water from the oceans, causing atmospheric circulation or convection. When the air reaches

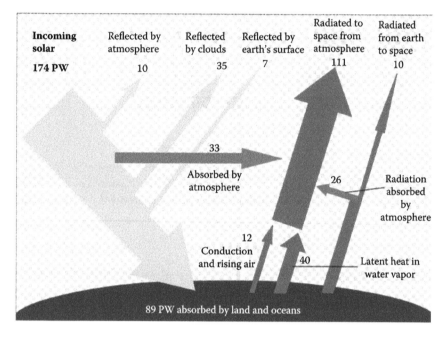

FIGURE 15.1
About half the incoming solar energy reaches Earth's surface. (From en.wikipedia.org/wiki/ Solar_energy.)

a certain altitude where the temperature is low enough, water vapor condenses into clouds, which rain onto Earth's surface, completing the water cycle. The latent heat of water condensation produces atmospheric phenomena such as wind and cyclones.

The total solar energy per year absorbed by Earth's atmosphere, oceans, and land masses is approximately 3,850,000 exajoules (EJ) as shown in Table 15.1 [1]. Depending on the geographical location, the closer a location is to the equator, the more solar energy is available there. The whole world electricity energy (56.7 EJ) is much less than one ten-thousandth of the available solar energy (3, 850, 000 EJ). This indicates that if we can effectively use the solar and wind energy, the energy requirement is not a problem.

TABLE 15.1

Yearly Solar Fluxes and Human Energy Consumption

Solar	3,850,000 EJ
Wind	2,250 EJ
Biomass	3,000 EJ
Primary energy use	487 EJ
Electricity	56.7 EJ

Source: en.wikipedia.org/wiki/Solar_energy.

15.2 Wind Turbine Energy Systems

The wind turbine is one of the most promising energy sources, and it has attracted much attraction in recent decades and penetrated utility systems deeply compared to other renewable sources [2–6]. Unfortunately, the output voltage and frequency of wind turbines are unstable as the wind speed is variable. These turbines are installed onshore or offshore, or sometimes as a wind farm where large numbers of turbines are installed and connected together.

Large-scale flowing air is called wind. Because of the sun, the wind always exists. Wind energy is from the sun and is a renewable energy. Figure 15.2 shows wind production. Atmospheric air circulates as in a boiler. The air becomes light at the equator, and heavy at the two poles. The wind flows day and night.

Wind energy has been used over the world from the ancient times; as an example, the windmill has been used for water pumping and grain grinding for over 3000 years. The wind turbine is modern equipment to convert the wind's dynamic energy into electrical energy. Two types of wind turbines are used: vertical axis and horizontal axis turbines [2–6].

15.2.1 Technical Features

The wind speed is uncertain and is presented as the Weibull probability density function (pdf) in Equation (15.1) or Rayliegh pdf with $k = 2$ and Figures 15.3 and 15.4 [2].

$$f(v) = \frac{k}{c}\left(\frac{v}{c}\right)^{k-1} e^{-\left(\frac{v}{c}\right)^2} \tag{15.1}$$

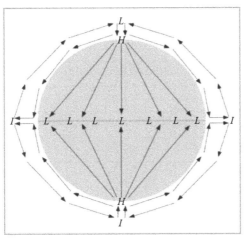

FIGURE 15.2
Wind production.

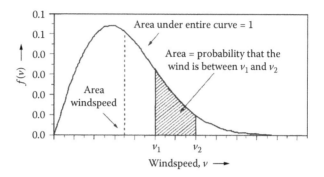

FIGURE 15.3
The Weibull probability density function (pdf).

In the above equation, k is the shape parameter, v is the wind speed, and c is the scale parameter, which can be 4, 6, 8, and so on depending on the particular location; for example, $c \approx 6$ at Singapore (see Figure 15.4).

The mass of air flows m in Figure 15.5 through a wind turbine is

$$m = A_1 v = A v_b = A_2 v_d \tag{15.2}$$

The power extracted by the turbine is

$$P = \frac{1}{2} m(v^2 - v_d^2) \tag{15.3}$$

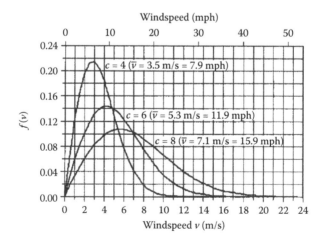

FIGURE 15.4
Rayleigh pdf.

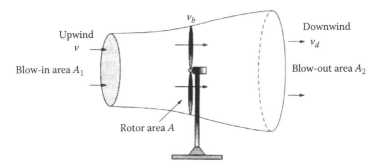

FIGURE 15.5
The mass of air flows through a wind turbine.

Since $m = \rho A v_b$ and

$$v_b \approx \frac{1}{2}(v + v_d) \tag{15.4}$$

$$P = \frac{1}{2}m(v^2 - v_d^2) \tag{15.5}$$

or

$$P = 0.5\rho A v^3 \left[\frac{1}{2}(1+\lambda)(1-\lambda^2) \right] \tag{15.6}$$

where ρ is the air density, A is the rotor area, v_b is the wind speed through the turbine, v is the blow-in wind speed, and v_d is the blow-out wind speed. To describe the power produced by the wind turbine, we use *Betz's law*:

$$P = 0.5\rho\pi R^2 v^3 C_p \tag{15.7}$$

where R is the radius of the windmill (or the length of the blade), and C_p is the power coefficient:

$$C_p(\lambda) = 0.5(1+\lambda)(1-\rho\lambda^2) \tag{15.8}$$

with $\lambda = \frac{v_d}{v}$ being the wind speed deduction ratio.
    The effect of temperature and pressure of air density is described by the gas state equation,

$$PV = nRT \implies \frac{n}{V} = \frac{P}{RT} \tag{15.9}$$

where n is the mass of air in mols; V is the volume of air in m³; P is the pressure in atm; R is the idea gas constant = 8.2056×10^{-5} in m³·atm·K⁻¹·mol⁻¹; and T is the absolute temperature in K.

The air density is given by

$$\rho = (n/V) \cdot M = \frac{P}{RT} \cdot M$$

where M is the molecular weight of air in kg/mol, 0.02897. Substituting M and R, we obtain

$$\rho = 353 \frac{P}{T} \tag{15.10}$$

The unit of ρ is kg/m³.

The atmospheric pressure is dependent on altitude. It is 1 atm at the sites at mean sea level. At sites above sea level, pressure is less; it can be shown as

$$\frac{dP}{dh} = -\rho \cdot g[(N/m^2)/m] \tag{15.11}$$

where h is the height above mean sea level, g is the gravitational acceleration constant (9.806 m/s²), and 1 atm = 1.01325×105 N/m². Therefore,

$$P = e^{-\frac{0.0341}{T}h} \tag{15.12}$$

The wind speed at a location varies with height h. The relation is

$$\frac{v}{v_0} = \left(\frac{h}{h_0}\right)^{\alpha} \tag{15.13}$$

where v is the wind speed at the height h, v_0 is the wind speed at the height h_0 (usually $h_0 = 10$ m), and α is the friction coefficient (see Table 15.2).

Usually, the wind turbine works in a certain range as shown in Figure 15.6 [4]. The wind speed change causes the output voltage and frequency to vary.

15.2.2 Design Example for Wind Turbine Power System

Figure 15.7 shows the block diagram of a wind turbine power system, which consists of a wind turbine, converters, and control subsystems.

TABLE 15.2

Friction Coefficient α for Various Terrains

Terrain Characteristics	Friction Coefficient α
Smooth hard ground, calm water	0.10
Tall grass on level ground	0.15
High crops, hedges, and shrubs	0.20
Wooded countryside, many trees	0.25
Small town with trees and shrubs	0.30
Large city with tall buildings	0.40

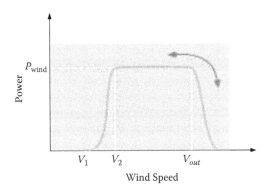

FIGURE 15.6
The wind speed range (power versus wind speed). (From Johnson, G. L. 1985. *Wind Energy Systems*. New Jersey: Prentice-Hall. With permission.)

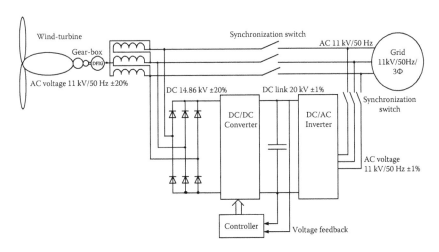

FIGURE 15.7
Block diagram of the wind turbine power system.

15.2.2.1 Design Example for Wind Turbine

The wind turbine feeds power to a 3-phase, 11 kV, 50 Hz grid through a wound-rotor induction generator operating with slip-power control. The configurations of the system are as follows [3–6]:

Induction generator:

Three-phase, 11 kV, 50 Hz, 4-pole, delta-connected

Per-phase magnetizing inductance referred to the stator = 7 H

Per-phase rotor winding resistance referred to the stator = 30 Ω

Stator to rotor turns ratio = 3:1

Gear box data: Power efficiency = 85%; speed ratio = 65

Site:

Altitude, 500 m above mean sea level

Average temperature, $30°C \approx 303°K$

Friction coefficient a of terrain is 0.15

Wind speed 7.79 m/s at a height of 10 m above ground.

Wind turbine:

Horizontal-axis, 2-blade

Diameter 50 m (R = 25 m)

Tower height 70 m

Efficiency = 45% at tip-speed ratio of 5.5.

Questions are to determine the following:

1. Slip of the generator
2. Mechanical power converted to electrical form
3. The magnitude, phase, and frequency of the phase voltage injected into the rotor by taking the stator terminal voltage as the reference phasor
4. Real and reactive power supplied by the rotor side converter
5. Real power supplied to the grid by assuming no losses in the converters and in the stator winding

Solution:

1. Slip of the generator

$$s = \frac{\omega_s - \omega_m}{\omega_s}; \quad \omega_s = \frac{4\pi f}{P} = \frac{4\pi * 50}{4} = 157.1 \, rad / s$$

Shaft speed w_m is directly decided by the wind speed. The wind speed at height of 10 m above the ground is given. Average wind speed is calculated at the turbine midpoint, that is, at the top of the tower

$$\frac{v}{v_0} = \left(\frac{h}{h_0}\right)^\alpha \Rightarrow \frac{v}{7.79} = \left(\frac{70}{10}\right)^{0.15} \Rightarrow v = 10.43\,m/s$$

Tip-speed ratio (TSR) = 5.5 = tip speed/wind speed

$$\omega_t \cdot \frac{D}{2} = v \cdot TSR \Rightarrow \omega_t = 10.43 * 5.5 * \frac{2}{50} = 2.295\,rad/s$$

Speed conversion through gear box:

$$\omega_m = 65\omega_t = 65 * 2.295 = 149.175\,rad/s$$

Therefore, the slip of the generator:

$$s = \frac{\omega_s - \omega_m}{\omega_s} = \frac{157.1 - 149.175}{157.1} = 0.05$$

2. Mechanical power converted to electrical form:

$$p_m = \eta_g P_t = \eta_g \eta_t P_w = \eta_g \eta_t \left(\frac{1}{2}\rho A v^3\right)$$

Hence

$$p_m = \frac{1}{2}\eta_g \eta_t \left(\frac{353}{T}e^{-\frac{0.0341h}{T}}\right)\frac{\pi D^2}{4}v^3$$

where T = 303 K; h = 500 +70 = 570 m; v = wind speed at turbine midpoint = 1.043 m/s; D = 50 m; η_g = efficiency of gearbox = 0.85; η_t = efficiency of turbine = 0.45. Substituting the data into the formula,

$$p_m = 465540\,W = 465.54\,kW$$

3. Rotor injected voltage

Delta-connected stator winding: $V_{phase} = V_{line} = 11000\,V$
We choose it as the reference phasor: $V_1 = 11000\angle 0°\,V$
Since there are no stator losses,

$$P_1 = P_g = \frac{P_m}{1-s} = \frac{-465540}{1-0.05} = -490042\,W$$

And

$$P_1 = 3V_1 I_1 \cos \Phi_1 \implies I_1 = \frac{-490042}{3 * 11000 * 1.0} = -14.85 \, A$$

$$I_m = V_1 / jX_m = \frac{11000}{2\pi * 50 * 7} = -j5 \, A$$

$$I_2 = I_1 - I_m = -14.85 + j5 \, A$$

$$V_1 = \frac{V_2}{s} + I_2 \frac{R_2}{s} \implies$$

$$\frac{V_2}{0.05} = V - (-14.85 + j5)\frac{30}{0.05} = 19910 - j3000$$

$$V_2 = 0.05(19910 - j3000) = 995.5 - j150 = 1006.7 \angle - 8.57° \, V$$

The rotor injected voltage is calculated using the turns ratio as

$$V_{1'} = \frac{1}{3}V_2 = 335.6 \angle - 8.57° \, V$$

The rotor injected voltage should have a frequency given by

$$f_2 = sf_1 = 0.05 * 50 = 2.5 \, Hz$$

4. Complex power absorbed by the rotor converter:

$$S_r = 3V_r I_r^* = 3V_2 I_2^* = 3 * 1006.7 \angle - 8.57° * (-14.85 - j5)$$

$$S_r = 3 * 1006.7 \angle - 8.57° * -15.67 \angle 18.6° = -47325 \angle - 10.04°$$

$$S_r = P_r + jQ_r = -46600 + j8250$$

$$P_r = -46.6 \, kW \quad and \quad Q_r = 8.25 \, kVAr$$

Real power and reactive power injected into the rotor winding by the rotor side converter are 46.6 kW and −8.25 kVAr.

5. Real power drawn from the grid:

$$P_{grid} = P_1 - P_r = -490042 + 46600 = -443442 \, W$$

This is also equal to the sum of the power converted to mechanical form (P_m) and the rotor current loss.

15.2.2.2 Design Example for Converters

The wind speed is unstable and changes from time to time in a certain speed range. The output voltage and frequency of the double-feed induction from time to time generator (DFIG) change by about ±20%. In order to transfer the unstable electrical energy generated from DFIG to the grid, we design our converter system as follows [3–6].

We assume the output voltage (e.g., line-to-line rms 11,000 V) and frequency (e.g., 50 Hz) of the double-feed induction generator (DFIG) change about ±20%, and the grid voltage (e.g., line-to-line rms 11,000 V) and frequency (e.g., 50 Hz) are very stable with ±1% variation. The converter's system design includes three parts: AC/DC rectifier, DC/DC converter, and DC/AC inverter, as shown in Figure 15.7. The AC/DC rectifier is an uncontrolled diode full-bridge rectifier. Its output is an unstable DC voltage of about 14.86 kV ±20%. The DC/DC converter is a boost type with closed-loop PI control. Its output voltage is 20 kV ±1% and is very stable. The DC/AC inverter is a VSI. Its output is a three-phase, 50 Hz, 11 kV (line-to-line rms).

15.2.2.3 Simulation Results

The simulation diagram is shown in Figure 15.8. The simulation results are shown in Figure 15.9. When the input voltage and frequency changed by 20%, both output voltage and frequency of the system remained stable.

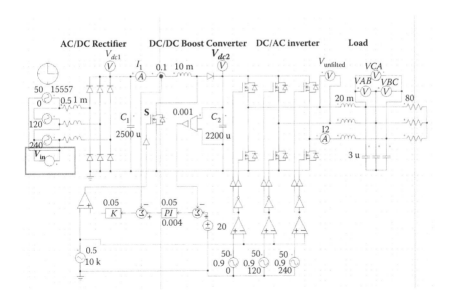

FIGURE 15.8
Simulation diagram of the wind turbine power system. (a) $V_{in} = 8.8$ kV (line-to-line, rms)/40 Hz (b) $V_{in} = 11$ kV (line-to-line, rms)/50 Hz (c) $V_{in} = 13.2$ kV (line-to-line, rms)/60 Hz

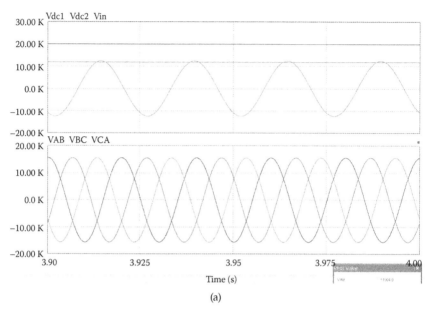

(a)

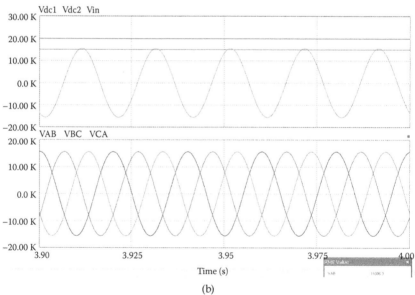

(b)

FIGURE 15.9
Simulation results of the wind turbine power system.

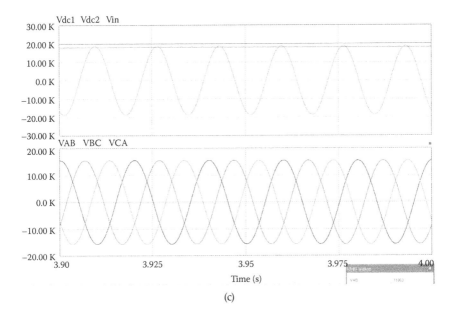

FIGURE 15.9 (continued)
Simulation results of the wind turbine power system.

15.3 Solar Panel Energy Systems

The sun offered sunlight and heat (with chemical effects) to Earth over millions of years, and this will continue for millions of years. The tremendous energy from the sun is thousands of times higher than the current total energy consumption of the world.

15.3.1 Technical Features

The sunlight changes from time to time. If the rated voltage of a solar panel is 186 V with a current of about 13 A, during a day it varies from 186 – 20% to 186 + 20%, that is, from 148.8 to 223.2 V. In order to convert this energy into the grid, we have to design appropriate power electronic circuits. The objectives are as follows:

1. To convert the unstable DC voltage to a stable DC voltage
2. To invert the stable DC voltage into 3 Φ AC voltage
3. To link the solar panel system to the main grid of 400 V/50 Hz/3 Φ

According to the above, the technical features are set as follows:

1. To match the grid data, we need an inverter to provide its output of 400 V/50 Hz/3 Φ with 1% variation.
2. To provide the inverter with output of 400 V/50 Hz/3 Φ, we need to offer a DC link voltage 700 V with 1% variation.
3. Since the input voltage is 186 V with ±20% variation, we need a high-voltage-transfer-gain DC/DC converter. The positive output super-lift Luo converter is selected.
4. To keep the link voltage at 700 V with 1% variation, we need a closed-loop control for the DC/DC converter.

We will briefly introduce each block of the system before beginning the system design.

15.3.2 P/O Super-Lift Luo Converter

The super-lift Luo converter is very good at high-voltage transformation and thus was used in the solar panel energy system. The positive output super-lift Luo converter [4–6] is shown in Figure 15.10. It consists of a switch S, an inductor L, two capacitors C_1 and C_2, two diodes D_1 and D_2, and a resistive load R. The input voltage is V_{in} and output voltage V_O, the switch frequency is f, the period $T = 1/f$, and the switch-on duty cycle is k. To avoid the parasitic effect, k is (0.1 – 0.9).

When switch S is on, the source voltage V_{in} charges the capacitor C_1 to V_{in}, and current flows through the inductor L. The inductor current increases by

$$\Delta I_L = \frac{V_{in}}{L} kT \tag{15.14}$$

When the switch S is off, the inductor current decreases with the applied voltage $(V_O - 2 V_{in})$. Therefore, the inductor current decrement is

$$\Delta I_L = \frac{V_O - 2V_{in}}{L}(1 - k)T \tag{15.15}$$

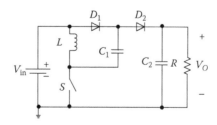

FIGURE 15.10
Positive output super-lift Luo converter.

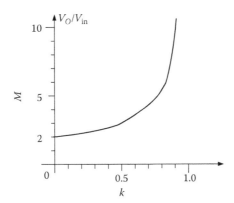

FIGURE 15.11
The voltage transfer gain *M* versus. the duty cycle *k*.

In the steady state, the inductor current increment must equal its decrement. Therefore, we obtain the voltage transfer gain *M* is

$$M = \frac{V_O}{V_{in}} = \frac{2-k}{1-k} \tag{15.16}$$

This voltage transfer gain is much higher than that of the boost converter and positive output Luo converter. When k is very small, the voltage transfer gain $M \approx 2$. When $k = 0.5$, the output voltage V_O is equal to $3 \times V_{in}$. The voltage transfer gain M versus the duty cycle k is shown in Figure 15.11.

In our system, $V_{in} = 186$ V and V_O required for the DC/AC inverter is 700 V. The voltage transfer gain M requested is 3.76; thus, the duty cycle $k = 0.638$. Since the input voltage varies from 148.8 to 223.2 V, the voltage transfer gain M and the duty cycle k change: $M = 3.136$–4.704, and $k = 0.532$–0.73. These values are very good for the given variation range.

15.3.3 Closed-Loop Control

The input voltage from the solar panel varies in the range of 148.8 to 223.2 V. In order to obtain a stable output voltage, we have to design a closed-loop control for the positive output super-lift Luo converter. To this end, a proportional plus integral (PI) controller is used for outer voltage loop control, and a proportional (P) controller for inner current loop control. The control block diagram is shown in Figure 15.12.

The output PWM signal is used to control the duty cycle k for the positive output super-lift Luo converter. The switching frequency is usually chosen in the range of 50–500 kHz. Since this is an automatic control, no k value need be preset.

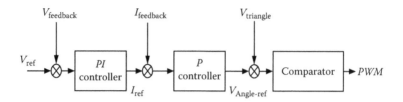

FIGURE 15.12
Double closed-loop controller.

15.3.4 PWM Inverter

The pulse width modulation technique is a popular method to implement DC/AC inversion technology. The pulse-width-modulated (PWM) voltage source inverter (VSI) introduced in Chapter 3 is used for this design. The three-phase full-bridge VSI is shown in Figure 15.13.

The triangular and modulating signals are shown in Figure 15.14.

There are two important modulation ratios for the PWM technique. We define the amplitude modulation ratio m_a as

$$m_a = \frac{V_{in-m}}{V_{tri-m}} \tag{15.17}$$

where V_{in-m} is the amplitude of the control (sine) waveform and V_{tri-m} is the amplitude of the triangle waveform. Usually, for nondistorted inversion the amplitude modulation ratio m_a is selected to be smaller than 1.0.

We also define the frequency modulation ratio m_f as

$$m_f = \frac{f_{tri-m}}{f_{in-m}} \tag{15.18}$$

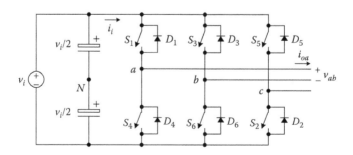

FIGURE 15.13
Three-phase full-bridge VSI.

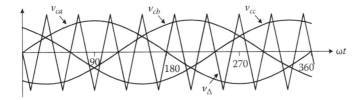

FIGURE 15.14
The triangular and modulating signals.

where f_{in-m} is the frequency of the control (sine) waveform, and f_{tri-m} is the frequency of the triangle waveform. Usually, for nondistorted inversion, the frequency modulation ratio m_f is selected to be greater than 21. The AC output voltage and current (each phase) are shown in Figure 15.15.

In order to produce the three-phase AC voltage to synchronize to the main grid voltage, we take the grid signals as the control signal.

15.3.5 System Design

After all the blocks are prepared, we can install our system. The block diagram is shown in Figure 15.16. The solar panel yields an input voltage of 186 V ± 20%. The DC/DC converter is the positive output super-lift Luo converter with double closed-loop control. Its output voltage is the DC link voltage with 700 V ±1%. Since the DC link voltage is quite stable, there is no need for any closed-loop control for the DC/AC voltage source inverter.

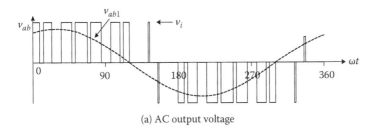

(a) AC output voltage

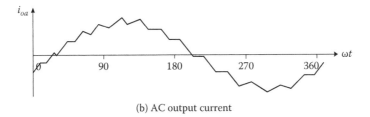

(b) AC output current

FIGURE 15.15
The AC output voltage and current (each phase).

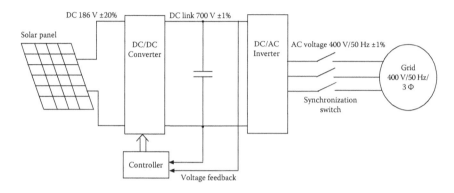

FIGURE 15.16
Block diagram of the solar panel power system.

Considering the synchronization, we use the grid voltage as the control signal of the VSI. Its output is a three-phase, 50 Hz, 400 V (line-to-line rms).

15.3.6 Simulation Results

The simulation diagram is shown in Figure 15.17 and the simulation results in Figure 15.18. Figure 15.18a shows the input voltage V_{in} is $186 - 20\% = 148.8$ V.

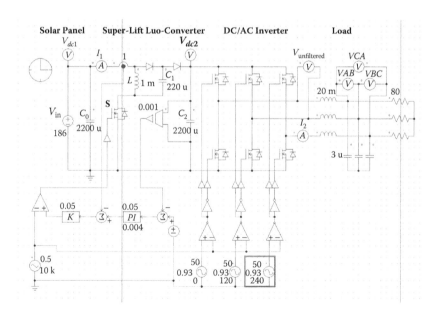

FIGURE 15.17
Simulation diagram of the solar panel power system. (a) $V_{in} = 148.8$ V, $V_{dc2} = 700$ V and $V_O = 400.155$ V/50 Hz/3 Φ. (b) $V_{in} = 186$ V, $V_{dc2} = 700$ V and $V_O = 399.906$ V/50 Hz/3 Φ. (c) $V_{in} = 223.2$ V, $V_{dc2} = 700$ V and $V_O = 400.321$ V/50 Hz/3 Φ.

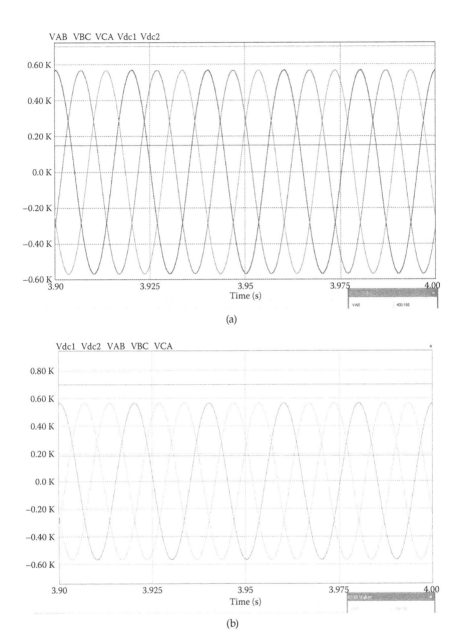

(a)

(b)

FIGURE 15.18
Simulation results of the wind turbine power system.

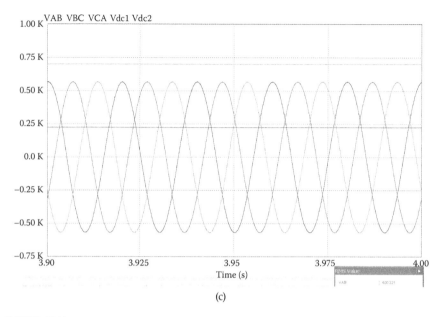

(c)

FIGURE 15.18
Simulation results of the wind turbine power system.

After the double closed-loop control, the output voltage of the P/O super-lift Luo converter V_{dc2} is 700 V. We then obtain V_O = 400.155 V/50 Hz/3 Φ after the DC/AC inverter. Figure 15.18b shows the input voltage V_{in} = 186 V, V_{dc2} = 700 V, and V_O = 399.906 V/50 Hz/3 Φ. Figure 15.18c shows the input voltage V_{in} = 186 V + 20% = 223.2 V, V_{dc2} = 700 V, and Vo = 400.321 V/50 Hz/3 Φ. In all cases, when the input voltage varies, the output voltage remains stable. The requirements of the application are thus satisfied.

References

1. en.wikipedia.org/wiki/Solar_energy
2. Masters, G. M. 2005. *Renewable and Efficient Electric Power Systems*. New York: John Wiley & Sons.
3. Ackermann, T. 2005. *Wind Power in Power Systems*. New York: John Wiley & Sons.
4. Johnson, G. L. 1985. *Wind Energy Systems*. New Jersey: Prentice-Hall.
5. Luo, F. L. and Ye, H. 2004. *Advanced DC/DC Converters*. Boca Raton, FL: CRC Press.
6. Luo F. L. 2012. *Lecture Notes on Renewable Energy Systems*. NTU Course EE4504.

Index